D1218715

Systematics as Cyberscience

Inside Technology

edited by Wiebe E. Bijker, W. Bernard Carlson, and Trevor Pinch

A list of books in the series appears on p. 293.

Systematics as Cyberscience

Computers, Change, and Continuity in Science

Christine Hine

The MIT Press
Cambridge, Massachusetts
London, England

For information about special quantity discounts, please email special_sales@mitpress.mit.edu

This book was set in Stone and Stone Sans by SPi Publisher Services, Pondicherry, India, and was printed and bound in the United States of America.

Library of Congress Cataloging-in-Publication Data

Hine, Christine.
Systematics as cyberscience : computers, change, and continuity in science / by Christine Hine.
p. cm.—(Inside technology series)
Includes bibliographical references and index.
ISBN 978-0-262-08371-3 (hardcover : alk. paper)
1. Biology—Classification—Data processing. 2. Information storage and retrieval systems—Biology. I. Title.

QH83.H56 2008
578.01'2—dc22 2007018944

10 9 8 7 6 5 4 3 2 1

For Dennis John Hine, 1926–2007

Contents

Acknowledgments

This project began way back in the early 1990s at the University of York, when a DPhil that was intended to focus on databases for biological nomenclature turned into a project focusing on a much wider scope of questions, concerning the way that biologists think about taxonomy. I am extremely grateful for those who at the time supported that shift in focus and enabled me to broaden my horizons. Special thanks are due to Michael Usher, to Michael Dadd and Georgina Mackenzie of BIOSIS, to Frank Bisby, and to Bob Allkin. Subsequently I have to thank Michael Mulkay for giving me the initial impetus to explore the sociology of science, and then Steve Woolgar for having the faith to back me in my efforts to change discipline.

My initial DPhil project was supported by a studentship from the Science and Engineering Research Council. More recently, I need to acknowledge the support of the Economic and Social Research Council for the three-year research fellowship (R000271262) which allowed me to return to my initial interests in taxonomy and to pursue them in the light of the current concern with e-science. I am also grateful for the encouragement of colleagues at Brunel University, and latterly at the University of Surrey. Numerous conferences and seminars have provided testing grounds for aspects of the argument contained in this book, and the feedback of colleagues at these events has been a source of invaluable stimulation. I am particularly mindful of the insights that collaborations with colleagues at the Virtual Knowledge Studio in Amsterdam have offered me.

The next set of acknowledgments I must make are to those who have informed and enlightened this recent phase of research, offering their time for interviews and guided tours, sharing with me their concerns and trusting me with their insights. I am very grateful to all those who have engaged with this research project, both face to face and in virtual space. The errors in this piece are mine, but much of its insight I owe to the contemporary biologists described within it who are also, along the way, extremely astute practical sociologists.

Finally, an acknowledgment to family is in order. During the time that I have been preparing this book two children, Esther and Isaac, have come along. I'm not not sure which is more challenging, birthing books or birthing babies. Thanks to Simon for making it possible to do both.

1 Introducing the Study of a Cyberscience

The connection between new technologies and social change is an issue of contemporary hope and fascination. We look to new technologies to produce desirable change and we fear undesirable effects, finding in computers a particularly potent stimulus for speculation. We watch with awe as changes that were neither anticipated nor desired come to pass, and we puzzle as structures we thought fragile turn out to endure. In this book I continue the exploration of the dynamics of change and continuity in relation to information and communication technologies (ICTs) that I began in a previous work, *Virtual Ethnography* (Hine 2000). In that book I looked at the social dynamics which made use of the Internet meaningful in the specific context of a media event in the late 1990s. I took an ethnographic perspective on change, looking at the ways that people made sense of their Internet activities in terms both of transformation and of reinforcement of the status quo. This time I turn to an area where the hopes and expectations of transformation through ICTs have been particularly intense of late: science. I deploy an ethnographically informed style of inquiry to explore the ways in which use of ICTs makes sense to a particular group of scientists. I range more widely in search of meaning-making practices, exploring multiple dimensions of the scientific discipline that inform the sense which participants make of ICTs. I also spend more time offline than previously, as I explore the ways that the virtual discipline makes sense to its practitioners and find it rooted in diverse facets of the existing structures, practices, and material culture. I find myself in territory where the idea of change is particularly politicized, where reflexivity is a highly developed practice, and where there is heightened consciousness of the importance of heritage.

This book is a sociological account of how one of the oldest branches of science turned itself into one of the newest and became a cyberscience. As the branch of biology concerned with naming organisms and exploring their relationships, taxonomy, or systematics,[1] has a heritage that reaches back for centuries. Recently it has renewed itself, thanks at least in part to the use of computers. This book explores the emergence of a virtual culture of taxonomy and investigates how that culture is entwined with the past traditions, the material culture, and the institutions and practices of the discipline. My aim was to find out how the relationship between ICTs and this branch of science came about, and to explore the dynamics of change and continuity that this relationship entails. It is no accident that I chose this particular area to focus upon. As well as researching the transition that taxonomy has made through visits to initiatives and projects, through observation of taxonomy online, offline, and in the published literature and through interviews with protagonists, I have also to a large extent lived the transition. It therefore seems appropriate to start this introduction on an autobiographical note. This will then bring me to a more conventional introduction to the focus of the book and the content of the individual chapters.

An Autobiographical Entry into the Field

For a long time I thought I wanted to study plants, and not people. The early 1980s found me, rather dazzled by the grandness of my new surroundings, entering the Botany Department at the University of Oxford as an undergraduate. My memories of the course have become hazy with time, and no doubt the parts that I enjoyed or disliked the most stand out disproportionately through the haze. I recall afternoons immersed in drawing exotic plants from the Botanic Gardens, and perched on a stool at a wooden bench, peering down a microscope at a lacy array of plant cells. I remember field trips around the Oxfordshire countryside, wading through bracken, squelching through bogs, and crawling through carpets of primroses, checking their flower centers for the position of stamens and stigma. Casting meter square quadrats on the ground to survey the flora and puzzling over identification keys to find names for obscure little scraps of moss became familiar activities. Even final exams took place partly in the woods, as we traipsed about surveying and making notes, on our honor not to meet up in the undergrowth and share answers with one another. I remember deep frustration at weeklong practical exercises in genetics, centrifuging, pipetting, and incubating for days, only to find that the vital solution had been thrown out by accident on day two and that for the rest of the week I had been carefully nurturing nothing at all. What I do not remember, from any stage of my short career as botanist, is using a computer. Information technology was certainly around, and I now know there were innovations in computing happening in that very building at around the time that I was there; but they were not a part of my consciousness as a botany student at the time.

Lately I have been back to the same botany department that once trained me, and while the fabric of the building is unchanged, many of the internal fittings have altered. Traditional wooden benches have largely given way to more modern office and laboratory fitments. There are, significantly, computers on many desks. Computing has clearly become integrated into the work of those now in the department, not necessarily as a notable or topical element, but simply as a part of getting the job done. The experience of being a botanist has undoubtedly changed too, although there is no doubt that as well as using computers, botanists also still wade through bracken, or bogs, or wherever their particular plants of interest live. Some things remain recognizable, but botany, or plant science as it is more often lately known, has had changes of substance and of image since I left, and among those changes has been a more intensive use of computers.

As an undergraduate botanist in the early 1980s I managed not to touch a computer for three years. As a sociologist at the beginning of the twenty-first century I cannot imagine working without one. Doing academic research without using a computer is now almost unthinkable, in the same way as it is all but inconceivable that any modern institution could work without information technology. Looking around me I can see my present-day colleagues accepting the computer in their working lives in a whole variety of ways. As a well-equipped sociology department we have routine access to computers on our desks, to word-processing, bibliographic, and statistical software, online databases, electronic journals, and access grid technologies for distributed collaboration. We maintain our Web presence, as a department and individually, as routinely as once we might have produced a printed prospectus or updated a curriculum vitae. Email, even to colleagues in the next office, is a usual part of an increasingly time-stretched and space-shifted academic existence. Whether searching for previously published work, communicating and collaborating with colleagues, recording data, calculating results, or writing an article, most researchers will depend on digital technologies in some form to support a project at all stages. The availability of these technologies opens up new avenues of research and imperceptibly smoothes the work of existing projects.

ICTs seem to have transformed the climate in which research is conducted to such an extent that researching credibly without them has become in a short space of time not simply more difficult, but almost impossible. Within the span of my career, working practices seem already to have changed dramatically. Some would say that we have only just started to find out what new possibilities are opening up. The applications I have listed tend to mirror existing activities, or are limited to reproducing tasks that are fully recognizable in terms of what went before. But use of ICTs could potentially go much deeper, transforming the nature of scientific work and communication to produce results not so recognizable in terms of former practice. Studying the integration of these technologies with science right now, as this book does, provides the chance to check up on that ongoing process and to consider what is happening at a point when we still have the chance to open the black box (Latour 1987) and unpack the social dynamics within which the structures being set in place make sense. We have an opportunity to work out what assumptions and interests are shaping the developments that are occurring. There is also a possibility to learn from experience, since different scientific disciplines could learn some valuable lessons from one another for their own futures, particularly as they become sites for explicit intervention through initiatives in e-science and cyberinfrastructure.

Much of the ICT in science has been introduced with the hope that it will improve the work that researchers do or make it more efficient. Most projects now come with a pressure for results in the shortest possible time, whether that be due to the constraints of a funding cycle, the race to outperform competing groups, the rival pressures of administrative and teaching tasks (themselves also increasingly computer intensive), or the urgency imposed by a pressing question of, say, environmental conservation or policy recommendation. Whatever the source of time pressure, tools to increase the efficiency of research have an immediate appeal. Along with efficient use of time and resources another factor to encourage use of computers is provided by the importance of accuracy in science. Here computerization holds the promise of automation, of excluding human error and enhancing reliability and reproducibility, and thus fits in well with values that science holds dear. Without ever setting out to make our work more computer intensive, we pick up individual tools that seem appealing and that we see our colleagues using effectively, and thus we find ourselves year-on-year relying on a wider and slightly different raft of technologies to support our work. Change is variable in pace and often incremental. Every department has its leaders who always know the latest tips, and its members who lag behind, needing to be encouraged or bullied to use a tool adopted by everyone else: individual approaches and experiences vary within an overall advancing trend of computer use.

On a policy level, the increase in use of ICTs in science has some obvious attractions. Digital publishing offers the possibility of rejuvenating an increasingly expensive formal scientific communication system, giving more researchers access to more publications, quickly, cheaply, and conveniently. Informal scientific communication, via email, newsgroup, and mailing list, promises to tie researchers into networks and keep them abreast of the latest developments. On a larger scale, databases and networking initiatives offer to make work more efficient and cost-effective by enabling the sharing of results between geographically separated groups of researchers, and reducing the duplication of effort. Unprecedented levels of computing power available to researchers seem to demand matching grand visions from programs of research and analysis previously unimaginable. These grand projects need organizational innovation, large-scale funding, and the political and individual will to imagine them and make them happen. They thrive on persuasive visions of how science ought to be and how new futures can be realized through new technologies.

Amid the pressures for efficiency and accuracy and the impetus to realize grand visions and undertake large scale data-sharing projects, computing therefore appears to have an automatic appeal for promoting change in desirable directions. We know already, however, that this kind of promise is rarely realized. Time and again complex social dynamics defeat brave promises like the paperless office (Sellen and Harper 2002) or predictions that teleworking would put an end to commuting to work (Gillespie and Richardson 2000). The apparently blanket transformative properties of ICTs turn into complex and diverse patterns of take-up and impact which are often counterintuitive (Woolgar 2002b). Efficiency is only one part of a story which will involve competing problem definitions, interests, and identities. In exploring the cumulative transformation of a cultural domain as a complex field of sociotechnical interactions, we need then to focus on the processes that bring transformation about. What are the dynamics of change? What factors influence the course of innovation? Whose voices influence the process, and how are the outcomes assessed? Who has a stake in recognizing change, or in identifying continuity? Excluding technology as a transformative agent in its own right enables one to start looking for other agencies and processes that enable cultures, including science, to change and to be recognized as having changed in conjunction with technologies.

In this book I will be addressing these questions through a focus on the processes by which ICTs have been introduced into one branch of science, systematics. The intention is to consider the social dynamics that surround the introduction of ICTs as captured in the questions above, and to track the identities, roles, institutions, and responsibilities that emerge as a result. I explore how particular uses of ICTs make sense to those involved, with a specific focus on the processes of imagining that enable participants to think about themselves, the territories they inhabit, the goals they strive for, and the capacities of the technologies they hope to employ. The questions raised by looking at the promises made for computerization in science become ones that deserve answers on a very local scale. In the conclusion I draw some generalizable lessons, for the study of ICTs and for the support of ICTs in science in particular, but these are bracketed by the specific nature of the social dynamics that surround and shape any particular instance we study. The value of a case study is in providing a deeper understanding of process, which in turn provides questions with which to interrogate other settings. The focus of this book is on the potential transformation of one field of science, but the aspiration is to use this approach to question some of the assumptions in current science policy and contribute to the sociology of science more broadly.

In order to map out further the scope and style of this study of transformation in biology, I will continue with the autobiographical story. After graduating as a botanist I found myself without a clear direction. Not knowing what I wanted to do led me, like so many others, to take a postgraduate qualification: the subject that I chose was of all things a course in computing. With a confidence borne of ignorance I enrolled myself in 1986 in a course designed to produce the hybrid computer-literate biologists who were just beginning to be in demand in the discipline. The MSc in Biological Computation at the University of York had been running for several years already, drawing on the expertise of staff from biology and from computer science. We trained in Fortran, Basic, and Pascal, in numerical and statistical methods, in the structuring of databases, and in the interfacing of laboratory instruments with computers. Clearly, then, my experiences as an undergraduate botanist were not mirrored throughout biology. I might have managed not to encounter computing as an undergraduate, but the contents of the biological computation course revealed some well-rehearsed ways in which the connection between biology and computers was being forged. In such courses the foundations of the discipline of bioinformatics, a hybrid of biology and computing, were being solidified. The hunch that computing would be a good career move for a biologist proved to be very well founded for many of my colleagues in the course: I often encounter them in key roles as I explore the contemporary uses of computing in biology.

My own career moves continued to be unplanned and not entirely conventional. After a year spent in a job processing data for species distribution maps, I returned to study for a doctorate, this time looking at the use of databases in biology. The idea was to see whether information systems could give a solution to a problem that was topical at the time among taxonomists, the "nomenclatural instability" caused by classificatory changes affecting the names of organisms. The complex chains of synonyms built up as classifications were revised were thought to be damaging the status of the taxonomy and the reputation of taxonomists. Instability of names was said to interfere with the work of other biologists who depend on classifications for utility and for scientific content. Databases were being proposed as solutions to give a user-friendly interface to complex systems of synonymy. I spent my time as a doctoral student grappling with database design issues, but also interviewing biologists about their use of taxonomy, and becoming immersed in the concerns of a taxonomic community passionate about its work and worried about its sustainability. I increasingly wanted to know what kind of solutions the community would find attractive and sustainable, and how far change might be a feasible proposition. I found myself interested not so much in developing computer systems myself, as in the sociological questions that arose from their deployment. It was thus, thanks to some helpful and open-minded sociologists, that at the beginning of the 1990s I found myself beginning a new career as a sociologist of science and technology.

Since leaving biology my sociological interests have ranged across a variety of areas far removed from taxonomic databases looking at the social processes that confer sense on particular technological solutions. The issue of the role of ICTs in knowledge production remains a live and practical concern for me as a sociologist as much as it is for the biologists I have been studying. In the sociology department, as in the contemporary botany department, computing and computer-mediated communication are used routinely, sometimes the topic of particular initiatives but more often an unremarked part of getting the job done. Our experience of change happens on different levels. Sometimes we consciously set out to do something new; at other times we may be jolted into a consciousness of change by reflection on how different our working style was just a few years ago. I have been conscious in writing this book that compared to the last time I have quite different resources available to me, as I now write accompanied by an always-on Internet connection and have an extensive personal library available to me through my bibliographic software. I wonder how my writing and my connection with my discipline may have changed with these alterations in my writing practices, and I strive to remember still to go to libraries and seek out the nondigital literature. I find myself keen to change in desirable directions, but suspicious of being changed by default.

The issue of changing science, and the role of ICTs in that process, is therefore something that interests me personally as well as theoretically, and it has become a stimulating place to pursue the interests in the social dynamics of meaning-making around ICTs that have concerned me as a sociologist of science and technology. Sciences and social sciences, like so many areas of contemporary life, seem to have undergone a quiet transformation in working practices, and a variety of initiatives seem to want to change them still further. In the next chapter I explore how in policy circles some aspects of the use of computers in science have become topical as sites for intervention in how science is done, while other kinds of computing work have been overlooked altogether. It is timely to reflect on the assumptions about technology that inform policy directions and routine work, and to think about the dynamics of continuity and change that encounters with new technologies occasion for science. Science and technology studies (STS) provides the tools for the critical perspective both on scientific practice and on the technology that this task requires.

Positioning a Case Study of a Cyberscience

The culture of contemporary science, including its adoption of ICTs, deserves documentation in its own right. There are also, however, more pragmatic rationales for carrying out this study, one of which is to contribute to understanding of ICTs as a site of policy intervention in contemporary science. To information scientists implementing ICT solutions in science, I offer a means to think about the dynamics of the situations in which their work intervenes. The book will, I hope, also be interesting to practitioners of the science that I discuss. I am sure that they will be infuriated by omissions and misunderstandings in my representation, but I hope that they will find their work described with respect and sympathy for their day-to-day experiences of getting their job done. Systematics has a strong tradition of self-examination, and this book, in offering new light on the structures and practices of the discipline, may fuel this tradition. To an audience within STS, I hope to contribute to understanding of the mechanisms by which scientific cultures change and are seen to change, and to demonstrate the interweaving of policy pronouncements and daily experience in the work of a scientific discipline. Finally, for anyone interested in the role of ICTs in contemporary society and concerned with how we can best tease out and explore the social implications of these technologies, I provide a case study that demonstrates the multidimensional cultural embedding of these technologies within a key area of social life over time. Cyberscience should give Internet research an interesting case study with which to work.

The term cyberscience needs some introduction. By using it I do not intend to indicate that there has necessarily been a radical shift or discontinuity in the ways in which science is practiced and communicated. Rather, the term is used as a qualitative indicator of the increasingly intimate relationship between scientific research and ICTs. As in many spheres of work, the use of these technologies has become inseparable from the doing of the work: the concept of "doing it by hand" has become meaningless. The term cyberscience demarcates this intimate connection as a site for examination. Just as cyberspace is used to mean a form of space realized through ICTs, so cyberscience implies the realization of science through those technologies. This realization of science includes both the representation of knowledge and the practice of research: cyberscience is not just about communicating science, but about ways of doing science as well. Just as cyberspace is turning out not to imply a replacement of real space, so cyberscience is unlikely, I would suggest, to displace or make irrelevant our existing notions of science. As the term cyberspace has been a rallying point to make apparent the reality of the experience of immersion in ICTs, so I would like to use the term cyberscience to mark as notable the extent to which ICTs have come to pervade science. Framing cyberscience in this way aims to topicalize ICTs so as to interrogate their use. This interrogation acts in two directions. In the first place, it questions assumptions that more and better technologies are needed for science. It becomes possible to ask what these technologies are supposed to do, for whom and in what context. Working in the opposite direction, there is purchase in looking for aspects of technology use that are deemed routine or unremarkable, and bringing into the spotlight their role in shaping practices and outcomes.

There is a methodological risk in gathering up all uses of ICTs in science as a single phenomenon as if their unity were a given. This class of technologies is a useful starting point for a study, since in the complex of computer-mediated communication, computation, data processing, and distributed computer networks we find the technologies upon which so many predictions of social transformation have been pinned. It should remain though an open question for ethnographic exploration whether this grouping of technologies has meaning for practitioners in their everyday work, or whether quite different distinctions and categories come to the fore. I do not, therefore, intend to labor the point about cyberscience too much or to suggest that the term will necessarily cover a unified phenomenon. Although the term cyberscience is useful, first as a handy shorthand for "the intensive use of ICTs within scientific research practices and knowledge representations," and second as a reminder to look both for the notable and the overlooked uses of these technologies in science, I do not intend to let the term get in the way of the analysis.

It may seem odd to base an examination of cyberscience on the discipline of biology. Biology is traditionally seen as the least technologically sophisticated and most craft based of the sciences. For example, Walsh et al. (2000) compared use of email within physics, experimental biology, sociology, and mathematics, and found that the biologists both had least experience using email and sent and received the fewest messages. However, there are some specific reasons why biology, and particularly the field of biology known as systematics or taxonomy, does form an interesting and instructive site to consider what cyberscience might entail. Possibly because of the traditional view of biology as nontechnical, new technologies have often been matters for debate within the field, and thus clear traces remain for a historical sociology of their introduction. Developments have been rapid, as I noted in my autobiographical sketch. Biology has achieved a considerable level of integration of research with information technology in a short space of time. Bioinformatics is now a recognized area, and while most prominent in genomics, this field has also had considerable influence within systematics. The process by which this happened is a fascinating complex of accommodations in research practices, roles, expertise, and technologies.

Within biology, systematics has particular value as a case study. Systematics has undergone long periods as an unfashionable and sometimes derided practice, but has emerged into the spotlight in the last decade with the growing political prominence of biodiversity conservation. As a discipline systematics has thus had to pay serious attention to its image, and use of technology has played a significant part. A wide range of ICTs has been incorporated into practice within systematics. Systematists routinely use email to communicate with one another, and mailing lists provide both for discipline-wide communications and for specialist groups to form within the discipline. Most systematists work in institutions such as natural history museums and botanic gardens, and much of their work revolves around the collections of specimens held in those institutions, which are used to develop ways of classifying and naming groups of organisms. We now have several decades of experience in developing computerized databases of specimen collections, increasingly now containing images as well as information from specimen labels, and publicly available via Web sites. Initiatives range in scale from the working databases of individual specialists to large-scale efforts to link the databases of several institutions. There are also high-profile efforts to develop user-friendly universal portals to distributed databases of biodiversity information, and to make available authoritative databases of all species names.

Systematics has recently been the subject of considerable attention on a global stage, as questions of biodiversity conservation have been debated, most significantly at the Rio Earth Summit in 1992 and in subsequent discussions around the international Convention on Biological Diversity signed at that summit. The particular spatial anomalies of systematics have been a prominent part of discussion. The distribution of specimen collections does not mirror the distribution of the organisms the specimens represent. Instead, major institutions hold concentrations of collections which have their own geography, often mirroring past patterns of colonization. Former colonies therefore find that the resources to understand their biological diversity are often far away in the institutions of their former colonizers. The Convention on Biological Diversity was formulated as a policy instrument to address these inequalities by enforcing the sharing of expertise and access to resources, and ICTs have often been seen as the practical route to achieving these goals. Without wishing to give too much of the story away, we could say that systematics proves to be a site where belief in technology as an agent of change, and as symbolic of a desire to change, has been particularly apparent.

Biology in general, then, is an appropriate site in which to look at the growth and significance of ICTs in scientific practice. Within biology, systematics provides a place to look at computer-mediated communication, at computerized analysis and databases, both as small-scale projects and large initiatives, as part of routine work and as a component of topical public debate. In this regard it is interestingly different from another obvious candidate for a case study, genomics. The Human Genome Project, as a high-profile international initiative, made intensive use of information technologies. Indeed, visions of technological possibility played a strong role in arguments that the initiative should be funded at all. Genome work is highly data intensive: sequencing of DNA produces masses of data, which can be meaningfully handled only by automated analysis, searching, and pattern matching. Procedures for submitting data to public databases are well established and integrated with the journals in the field. If one is looking for a model of cyberscience, then genomics would seem an obvious place to start.

There are reasons, however, why genomics is not the most useful case study for the particular purposes of this book. I wanted to be able to trace the changing expectations of computers over time as a discipline developed, and to look at the ways in which the technology was implicated in the working practices of the discipline and in the relationships of the discipline with its users. In this respect, a case study as driven by visions of large-scale computing as the Human Genome Project could be counterproductive. All case studies are, of course, unique. That is their strength as well as their weakness. I hoped, however, in the choice of systematics to have a case study that was continuous in some recognizable ways with the experience of other disciplines, as well as being interesting for its distinctiveness. In the conclusion I return to the question of general issues that may be drawn from the case study of systematics, and specifically this question of the comparisons with genome research, which as we see in chapter 3 is a live issue for participants as well as commentators.

Outline of the Book

In the remainder of this introduction I will describe the content of the book chapter by chapter, in order to further map out the approach and give a taste of the kind of insights that it provides for. Chapter 2 examines the technological focus of cyberscience, exploring first what claims are made for ICTs in science, and second what roots these ways of thinking have in our beliefs about these technologies more broadly. It appears that, through notions such as e-science and the collaboratory, high-profile ICTs have attracted policy interests in shaping scientific practice in particular directions. As a policy tool ICTs have an appeal as ways of promoting efficiency, data sharing, collaboration and speed. Viewed from a sociological perspective, however, these claims appear somewhat problematic. Not only may the perceived benefits not be realized, but other unpredictable consequences may transpire.

A critical approach to the expectations that shape policy interest in ICTs for science turns attention onto the processes through which technologies are adopted and interpreted by their users. Sociology of science and technology provides ways of thinking through the social dynamics of these processes of adoption. From studies of technological change in organizations, the concept of "technology in practice" (Orlikowski 2000) offers a route to thinking about the situatedness and specificity of any particular implementation of a technology. From sociology of science, the idea of co-construction provides a way to articulate the intricate relationship between the tools researchers use and the goals they aspire to achieve (Clarke and Fujimura 1992). Taken together, these approaches suggest an exploration of the ways in which scientific disciplines make ICTs their own, and, in the process, articulate new versions of themselves and their aspirations.

Taking the introduction of ICTs as a moment of co-construction provides a powerful rationale for an ethnographically informed investigation, looking for the practices within which particular technological solutions are made to make sense. ICTs are, at least by reputation, thoroughly troubling to spatial organization. It is precisely because they are thought to alter social relations in time and space that we find them so interesting: it therefore seems perverse to tackle understanding them through a spatially focused ethnography that looks at a laboratory (or an equivalent bounded location, such as an herbarium or museum). If we are to retain the possibility of discovering new organizational forms in science, we need to adopt a methodological approach that remains open to finding and exploring such forms. In an area as politically charged and subject to open intervention as ICTs, it also seems important to look deliberately beyond the immediate local environment rather than waiting for the global to emerge in the local in traditional ethnographic style. My case study of ICTs in systematics therefore employed a combination of multi-sited ethnography, historical review, and textual analysis across a series of settings. Building on the virtual ethnographic approach (Hine 2000), which insists that virtual settings are both cultural sites in their own right and cultural artifacts subject to ongoing processes of interpretation, the case study explored online landscapes as they made sense within diverse contextualizing frameworks. The findings of this research are arranged into the four chapters that follow, examining in turn the policy perspective, the material culture, the communication system, and the institutional/spatial form of systematics as they relate to the use of ICTs.

Chapter 3 begins the examination of the case study of systematics by both setting the scene for understanding contemporary systematics and exploring how ICTs are imagined in policy commentaries on the discipline. The data for this chapter are drawn largely from a 2002 policy report, produced by the UK's House of Lords Select Committee on Science and Technology, which reviewed the current state of the discipline and recommended that it should take steps to become a thoroughly Web-based discipline. The chapter examines the reasoning presented in support of this recommendation, and explores the themes that arise in the report and the evidence from stakeholders on which the report is based. Themes of importance include: the relationship between material and virtual specimens, and the audiences for their availability; the complex political geographies of systematics; the balance between automation and expertise; and the role of evocative objects and the branding of initiatives. Historically, these themes can be seen to have varied as the concerns of the discipline and its political context have altered. This chapter begins the work of establishing that ICTs in systematics are thoroughly shaped by the context of the discipline that employs them, which in turn consists of a highly distinctive culture in terms of its organizational forms, funding structures, working practices, and material culture. The introduction of ICTs into systematics proves to be a site for imagining the present, past, and future of the discipline.

A further perspective on the mutual articulation of ICTs and systematics is provided by looking at the way the discipline works with objects to produce its distinctive ways of exploring the natural world. Chapter 4 focuses on this material culture of systematics and investigates where the developing virtual culture supplements, replaces, and builds on that material culture. It emerges that such virtual culture as has developed is portrayed by systematists as thoroughly entwined with the beliefs and practices that surround their material culture. The discipline is founded on ordered collections of physical specimens as a representation of the diversity of the natural world. ICTs are providing new ways of ordering specimens and making them available to traditional audiences and wider publics. Virtual technologies allow for the material culture through which objects are collected, ordered, and displayed to be experienced in new ways. Processes of co-construction dominate as visions for virtual technologies are articulated and in turn provide new ways of understanding and specifying the qualities of material specimens.

In chapter 3, the role of ICTs in systematics is explored through analysis of public, high-profile commentary. Such talk is, of course, limited in the extent to which it reflects the daily experience of those working in the field. Even those practicing taxonomists who gave the evidence on which the report analyzed in chapter 3 was based were clearly speaking very consciously in a public arena. In chapter 4, communication practices that underpin and interpret the virtualization of taxonomy's material culture are explored in passing. Chapter 5 turns attention directly to communication practices, to ask how these have altered and been respecified in light of new technologies, and with what significance for the practices and outcomes of taxonomy. Having introduced the role of computer-mediated communication in developing understandings of the formal taxonomic communication system, the main focus of chapter 5 is a mailing list: a form of communication that is still public but in a much less formal and politically charged sense than the academic literature or the report that forms the basis for chapter 3.

A mailing list for taxonomists provides material for assessing the role of ICTs in the working practices of the discipline, in two senses. In the first sense, the existence of the list speaks to the use of the technology by members, and the membership of the list provides some indication of the extent to which this technology has become a routine way for taxonomists to communicate with one another. In a second sense, the content of the messages provides material to consider how much, and in what senses, use of ICTs is topical within the discipline. The list is frequently used as a way of announcing new products, which often take the form of digital technologies. The list attests both to the importance of these technologies and to the emphasis on publicizing one's products in contemporary systematics. The mailing list gives a perspective on taxonomy that contrasts with the wholesale shift toward Web-based science advocated by the Select Committee. It reveals a highly self-conscious and reflexive discipline, which is able to use the list to realize itself as a community, and yet maintains a very individualized set of goals and priorities, revolving around institutional affiliations and taxonomic groups.

The mailing list analyzed in chapter 5 was frequently used to publicize initiatives and products. Chapter 6 shifts focus to examine the institutional locations within which these initiatives arise and to consider how far ICTs are restructuring systematics on this level. It has become routine to use computerized analysis of some kind in classification, and curatorial work also routinely deploys databases. This has so far been achieved without significant change to the experience, location, and institutional organization of work in systematics. Large-scale database initiatives, however, involve more explicit and deliberate social innovation. The work of database construction involves negotiations to draw appropriate boundaries between taxonomy and its publics and to integrate the work of database construction and the production of data into existing working practices. Existing institutions still have a strong role to play in systematics, not least because they house the collections of material specimens on which the discipline is based. Systematics institutions are expected to make their resources accessible, but also have to maintain their distinctive identities and their fundability. ICTs provide ways of managing this tension, along with a reflexive opportunity to reconsider the status and importance of relations with users. The implementation of access initiatives provides a new occasion for thinking through what material specimens are used for and by whom, and for working out the relative status of material and virtual specimens and products. At the same time, individuals within institutions are conscious of career pressures, and initiatives are being carefully designed to give due credit, maintaining the existing emphasis within taxonomy on the reputation of individuals. This chapter demonstrates the complex dynamics between transformation and continuity that prevail on the institutional level.

Chapters 3 through 6 explore the uses of ICTs in systematics from a set of different perspectives: the policy domain, the material culture, the communication system, and the institutionally located set of working practices. Although treated separately, these ways of understanding the discipline are clearly thoroughly interwoven with one another. To have left out any one would be to have omitted a significant part of the environment of contemporary systematics. Including all of these perspectives allows us to see how ICTs are involved in systematics not just as isolated initiatives or as a different way of doing work, but as highly influential sites for imagining what the discipline is across all levels. Chapter 7 pulls together the implications of these observations from a policy perspective and for sociology of science and technology. Contemporary systematics is ICT saturated, but in a way that is so thoroughly shaped by the current concerns of the discipline that it becomes difficult to talk about the effects of ICTs as an agent of change. Instead, ICTs have been a useful resource for systematics to realize itself in the current climate, through the development of very distinctive manifestations of ICTs deemed appropriate to its own concerns. Understandings of ICTs for systematics are thoroughly situated in the detail of the current priorities, political status, and funding climate of the discipline but are also connected with imaginings of the potential of ICTs in other domains. The introduction of ICTs emerges as a therapeutic moment for systematics, involving co-constructions of the past, present, and future of the discipline, while also imagining change for the discipline and the territory within which that change is to occur.

To return to where we started, then, it is clearly problematic to talk about ICTs as ways of making science more efficient. It is true that database projects are providing unprecedented access to taxonomic data. Exciting initiatives are being undertaken by technically savvy and often idealistically motivated individuals to address persistent inequalities in availability of resources for taxonomic work. It is not clear, however, that direct benefits of such access are thoroughly understood, or indeed are always a significant driver behind the projects in the first place. A suitable political and funding climate can provide ample justification for projects to go ahead without any direct evidence as to benefits or any identifiable bodies of active users. Policy interventions which intend to use ICTs as a means to promote change in the science system therefore need to proceed with caution. The upshots may be highly specific to the political status, funding concerns, existing organizations, and working practices of a given discipline. It cannot be guaranteed that efficiency gains, as these are normally understood, will result.

This book is likely to offer the most to people already interested in science, but it also clearly has theoretical and methodological messages for the study of ICTs as cultural artifacts more broadly. The theoretical issues with wider significance than for science studies alone revolve around ICTs as resources for the reflexive reshaping of areas of practice. What people do with ICTs depends on understandings of both their functional qualities and their cultural meanings. What people do with these technologies is thus embedded in their daily understandings of goals and how these are to be achieved. At the same time, our understandings of what is to be done on the everyday level are thoroughly infused with a sense of the environment, both technical and cultural, in which we are working, and our sense of possibilities is shaped by the technologies that are available. An understanding of the development of ICTs in any area of social life can usefully draw on investigation across multiple facets of that area of life, including the details of daily practice and the occasions of public commentary and policy intervention.

My choice of site for the case study in this book was thoroughly shaped by my biography, and I felt it important that this introduction should have a strong biographical feel. It would be disingenuous to have given the impression that I came to systematics as any kind of distanced or impartial observer. From the outset I was not inclined to accept the stereotype of systematics as a dry pursuit, practiced by people cut off from the rest of the world, suffering, as Baroness Walmsley described it, "from an image of anorak-clad scientists poring over disintegrating specimens in dusty archives" (Hansard 2002: col. 921). I cared about the depth of history that systematics carried with it, and loved the fact that this was a branch of science that celebrated its past. At the same time, I knew it was a living discipline, and was convinced how important it would prove in the struggle to preserve biodiversity. I still, though, encountered surprises in the interviews and visits that I undertook for this book. Time and again I was taken aback by the political savvy, deep engagement, technological sophistication, and passionate will to explore new ways of working and to engage with global concerns that I encountered. I have portrayed these people as I found them, and have found my initial sympathies turned into a much increased respect. The mobile and connective approach which I adopted has allowed me to appreciate the complex world in which contemporary systematists work. I hope that they will see this book as complementary to their concerns.

2 Science, ICTs, and the Imagining of Change

Contemporary culture largely believes in the potential of ICTs to effect change, and science participates in and feeds this broader current of expectation. In particular, ICTs have been seized upon within the policy community as a means to promote desirable changes in scientific practices and aspirations. This book focuses on the social dynamics that produce the contemporary engagement between science and ICTs and examines the structures and practices that result by looking at the experience of one discipline. This provides ultimately a means to discuss the potential of ICTs as a lever of change and to examine the benefits and drawbacks of relying on these technologies to effect transformations. In the introductory chapter I took a largely autobiographical approach to establishing this focus of interest, telling my own story of the ways in which both my original discipline and my new one seemed to be changing in the face of developments in computing and communication. In this chapter I develop a more detailed rationale for a particular ethnographically informed approach to the study of ICTs and change in science, which moves away from an autobiographical account of a sense of change toward an understanding of social dynamics rooted in the sociology of science and technology.

This chapter divides into five sections, beginning with an examination of the relationship between science and ICTs as expressed within the policy domain. The role of ICTs within policy discourse, as a focus of expectation in relation to scientific practices, is juxtaposed with the more skeptical approach from STS. In this field, while "cyberscience" marks an area for critical attention, there is considerable caution about the likelihood of wholesale transformation. This critical perspective builds on the traditions both of sociology of technology and of sociology of science. In the three sections that follow this initial survey of policy discourse and STS, I formulate some elements for a critical STS-informed analysis of a cyberscience in the making, looking at literature on technologies and organizational change before moving to science specifically, examining first the technologies of scientific instrumentation and second the communication practices of science. In the final section of the chapter I draw these strands together into a methodological rationale for the case study I carried out in systematics.

Computers and Aspirations for Scientific Change

Computing has deep roots in academic science and engineering. This is so in the fairly obvious sense that the computer is a piece of technology developed through the application of scientific and engineering knowledge and practice. It is also true in the sense that much of the initial development of the computer and many subsequent advances in software and hardware happened at universities and occurred in response to problems formulated by science and engineering disciplines. The capacity of computers to perform calculations on a scale and at a speed unattainable by humans made particular sense within academic disciplines that relied on mathematical operations. For example, the Manchester Mark 1, among the first recognizable modern computers with the ability to run stored programs, was developed at the University of Manchester and used in scientific research within the university in 1949 (Ceruzzi 2003). Coming closer to the present, the rationale driving the development of computers in science now includes a lot more than this initial focus on mathematical calculation. For example, the initial formulation of the World Wide Web was concerned with the perceived need for communication among dispersed scientific research groups (Berners-Lee 2000). New computing technologies have both come from science and promised to change how research is done and the kinds of problem that it can address. As Aborn put it in 1988, "Science is on the verge of being transformed by one of its own prodigies—information technology" (Aborn 1988: 10).

The history of computing is thus interwoven with the history of science, and problems in science have provided a significant stimulus to computing developments that have often then been adopted more widely. Such is the extent and the complexity of the interaction between computers and science that it would be futile to attempt a comprehensive summary overview that disentangled the two. Agar (2006) begins the job of examining instances where computers have been credited for enabling new directions in science, and finds that the direction of influence is often unclear, or runs in the opposite of expectations. In this section I will therefore not survey the whole territory, but will be looking specifically for places where the use of computers in science becomes a site where intervention is proposed.

This provides the material for an exploration of some of the expectations that prevail around computers in science and the versions of technology and scientific practice that they formulate. In the final part of this section I then look at the ways that STS has faced up to the aura of expectation that surrounds ICTs in science, finding in it both assumptions to criticize and opportunities to engage in debate.

One of my favorite starting points for thinking about the claims made for ICTs in science is a speech given by the late Robert Maxwell (1990). The "media mogul" spoke on the occasion of the annual British Library Dainton lecture of the possibilities that information technology held as a way of making communication of scientific information more efficient, by which he meant it would be both quicker and cheaper than traditional means. As owner of numerous publishing interests and responsible for a suite of scientific journals as well as online information services, Maxwell clearly was not speaking from a neutral position. He expressed, however, a common way of thinking that associates information technology with communicative efficiency. This way of thinking has also been popular in policy forums: a 1998 Academia Europaea conference considered the role of electronic communication in research in Europe (Meadows and Böcker 1998), a European Technology Assessment working group considered the issue of how European science could be transformed through ICTs (ETAN Expert Working Group 1999), and a series of three OECD conferences and reports in 1996, 1998, and 2000 discussed "The Global Research Village" focusing on the potential of ICTs within the science system (OECD 2000). A series of reports in the United States has explored grand visions for ICT-enabled science (as described by Vann and Bowker 2006). The claims for ICTs in science in these contexts include the idea that they make the communication that underpins science easier, quicker, and cheaper.

Casting electronic communication as fast and cheap suggests that there is an existing medium that is slow and expensive. This contrasting scenario is often provided by traditional journal publishing. Increasing numbers of journals make it difficult for libraries to fund subscriptions, while decreasing subscriber numbers make journals less profitable for publishers. It seems self-evident, from this perspective, that Maxwell's (1990) vision of a future of electronic scientific communication is the way forward, and, as the extensive bibliography assembled by Nentwich (1999) suggests, there has been much speculation on the likely direction and rate of change. There have certainly been notable successes in electronic publishing within particular scientific cultures, such as the tradition of electronic preprints in the physics community (Merz 2006). On the whole, however, the traditional journal system is proving slow to change. Many journals are now available online, but few wholly online prestigious journals exist as yet. Nentwich (2001b) suggests that electronic communications for science may provide a new phase of decommodification in which information is increasingly distributed by means that bypass traditional publishers, but it is by no means clear that a universal and wholesale shift is underway (Nentwich 2006), and even if electronic publishing is the way forward there is a wide variety of models under which it could operate (Wellcome Trust 2003, 2004). The image of electronic communications as quick and cheap compared to paper-based media has driven an influential strand of thinking on the role of computers in science. However, the vision of universal electronic communication has yet to be realized, and it underplays the wide diversity of practices that prevail in both conventional publishing and electronic alternatives.

The need to support informal means of communication among scientists has also been prominent in thoughts about ICTs in science. Matzat (1998) suggested that electronic networks might be a way to break away from the "invisible colleges" that can exclude marginalized scientists. In the European context, ICTs for research have the attraction of promising to make communication between dispersed researchers seamless, fostering the development of a distinct European research area. Among the factors needed to promote the development of this area, electronic networks were identified as having an unrealized potential. The European Commission's statement on the development of the European research area included the following recommendation:

> To increase the productivity of European research while helping to structure collaboration on a continental scale action will have to be taken in this context to encourage the use of electronic networks in the various fields of research in European as well as national research programmes: development of databases and access to advanced Internet services; promotion of the production of multimedia content and interactive uses; support for new forms of electronic collaboration of researchers ahead of the emergence of real "virtual research institutes." (Commission of the European Communities 2000: 11)

The European vision sees electronic communications as a means to overcome the spatial distribution and the national segregation of research within its domain.

As a vision for spatially dispersed scientists working together, the notion of telescience has enjoyed some currency, particularly within space science. As originally formulated by NASA telescience was conceived to allow dispersed teams of earth-based scientists to control instruments and conduct experiments remotely in space, although the versions articulated by Aborn (1988) and by Lievrouw and Carley (1990) focus less on access to instrumentation and more on enhanced interpersonal interaction between geographically dispersed scientists. The WITS initiative (Backes, Tso, and Tharp 1999) enrolled the public, inviting them to use similar tools via the Web to participate in their own simulated versions of planetary landing explorations: a hierarchy of user types maintained distinctions between the levels of access and control allowed to different users. The scientists involved may or may not be colocated: in fact, since teams would generally be assembled from many different institutions they might well be working from different places. They are, however, inevitably distanced from the instruments and experimental materials they wish to work with. The major telescience applications developed in a space science context focus on problems of providing appropriate communication structures and real-time control to allow scientists to plan and to work with apparatus, and to provide interfaces that give appropriate access both to instrument controls and various forms of results and environmental monitoring. The model of telescience has subsequently been applied outside space science: similar approaches have been tried in other fields which rely on dispersed groups using data-intensive and expensive instrumentation, such as electron tomography (Lee et al. 2003).

The notion of the collaboratory has lately tended to replace telescience, offering a generalized vision of infrastructures to enable scientists to work together while apart, exchanging information and solving problems in shared online spaces. The collaboratory is portrayed as a new organizational form for the conduct of scientific research (Finholt and Olson 1997). Wulf (1993) described the collaboratory as an opportunity to use computers "to further leverage the entire scientific enterprise." While computers had become familiar as ways of carrying out analysis and mathematical modeling, new technological prospects offered the chance for impacts much more broadly across scientific disciplines. The vision that Wulf articulated, since adopted by many practical attempts to implement shared electronic working spaces for sciences, revolved around the provision of advanced databases, access to remote instruments, technologies specifically designed to help scientists collaborate, fusion of data from different sources, and aids to mathematical modeling. This is, therefore, a particular model of science as a collective activity, as data- and computation-intensive and as based on the need for sharing of information not just at the publishing stage, but throughout the process of research. The vision of the collaboratory as expounded by Kouzes, Myers, and Wulf (1996) is a thoroughly utopian one.

The authors are not convinced that the technology is currently in place fully to realize their dream, nor do they believe that the various psychosocial barriers to collaboration that they identify have been overcome. They present a vision to drive the development of collaboratories in order to allow scientists to address increasingly complex scientific problems.

Vann and Bowker (2006) argue that a focus on interdisciplinarity and collaboration was a distinctive development in US visions for the role of computing in science that emerged in the late 1980s and 1990s. They pinpoint the importance of Wulf's thinking in cementing the idea that the way forward for science was in addressing complex problems, and that to do this scientists would have to be brought together both spatially and across traditional disciplinary divides. Recent thinking in the UK on the application of ICTs in science builds on the idea of computing technology as a means to allow exploration of previously inaccessible areas of science, focusing on both computation and collaboration. The development of computers in the first instance was driven by the vision of making calculations more quickly and accurately than the human brain. This aspect of computing technology today takes on new connotations as masses of data for analysis accumulate from "Big Science" projects such as particle physics or industrialized genome sequencing. Current visions of computing needs in science encompass the need for a new scale of analytic power. The e-science initiative in the UK, funded by the Office of Science and Technology across the Research Councils, articulates the latest version of the promise of information technology for science:

> What is meant by e-Science? In the future, e-Science will refer to the large scale science that will increasingly be carried out through distributed global collaborations enabled by the Internet. Typically, a feature of such collaborative scientific enterprises is that they will require access to very large data collections, very large scale computing resources and high performance visualisation back to the individual user scientists. (UK Research Councils e-Science Core Programme n.d.)

This version of ICTs in science explicitly attends to a change in the scale of the science that can be done. The technologies are to be used not simply to make science cheaper or quicker, but also significantly to alter the problems that can be addressed.

It is important to note the origin of the two influential concepts, the collaboratory and e-science. Wulf, when he coined the term collaboratory, was Assistant Director at the US National Science Foundation (Kouzes, Myers, and Wulf 1996). The term e-science is attributed to Dr. John Taylor, Director General of the Research Councils in the Office of Science and Technology in the UK (Hey and Trefethen 2002). In both cases, then, a vision is articulated at a highly influential level in research policy and funding. ICTs appear as potential agents of change that provide an alternative to funding initiatives as a means of promoting shifts in scientific direction and organization. While Bud and Cozzens (1992) suggest that the usual policy interest in scientific instruments is how to respond to requests from scientists for more money, this particular vision turns matters around. Here policy is proposing, as much as or more than scientists are asking. Policy makers are developing visions for ideal future forms of scientific practice and proposing to support technological development to enable scientists to realize them. Looking outside science, Sandvig (2003) notes that forms of technological utopianism have become part of the mundane reality of public policy. As Sandvig points out, utopian talk is a way of criticizing the present. Sandvig (2003) suggests that casting the problem as one of finding the appropriate technology distances the policy pronouncement both from the critique and from responsibility for implementing the solution. If change fails to happen, it will be the technology and not the policy that is deemed to have failed.

Sandvig's observation makes clear that it is important to look both at the content of expectations about the relationship between technologies and social change and at who is making these claims on whose behalf. Brown and Michael (2003) make a case for a "sociology of expectations," focusing on understanding variations in predictions of the future and the diverse constituencies who create and use them. Proposals on a policy level for the introduction of new technologies can encompass visions for the future and critiques of the status quo and can include some constituencies while alienating others. This observation provides an important factor in outlining the scope of the case study within this book. It will be important to recognize that change is something in which a variety of communities have a stake. This is a territory in which change is politicized, and where there are parties who aim both to promote change in particular directions and to present change as having occurred. It would be oversimplistic therefore to design the case study in this book as an investigation of whether scientific culture has changed in the face of new technologies. It is as important to develop a consciousness of who is making claims about change and continuity, under what circumstances, and for which audiences.

This kind of critical and intrinsically social perspective on the dynamics of technological innovation in science is rooted in the heritage of STS. There have been suggestions that the coming of intensive ICT use in science raises interesting questions for STS, forming a sphere of concern sometimes captured in the term "cyberscience." This term was employed by Wouters in a call for papers for a session of the Society for Social Studies of Science/European Association for Studies of Science and Technology conference to be held in Bielefeld, Germany in 1996:

The session "Cyberscience" of this meeting will focus on the emerging new phenomenon of "cyberscience." The fast development of computer networks, the Internet, as well as electronic communication in science, social science and the humanities has many consequences, of which most are still unmapped and unexplored. This session wishes to contribute to the understanding of the implications this convergence of cyberspace and the sciences might have.

The following issues are on the agenda:

- the electronic scientific journal: is the printed paper disappearing?
- computer tools in science, especially related to networking
- distributed databases and their consequences for knowledge production
- issues of proprietorship in cyberspace: who owns the knowledge?
- the role of electronic archives
- networked management of research
- on-line evaluation of science
- the new virtual (even more invisible?) colleges
- networking and community formation in science
- the "virtual scientific instrument"
- new forms of electronic peer review
- desktop and on-line scientometrics
- gender issues in cyberscience
- theoretical perspectives of cyberscience studies
- does the net alter daily practice in scientific research?
- the dynamic document: a new form of science reporting?
- the creation of multi-media forms in science
- who will publish in cyberscience?—the struggle between the publisher and the scientist
- science performed by robots
- how do computer networks and information technologies shape themselves as well as our knowledge of them?
- science, technology and bureaucracy in cyberspace
- science policy: is networking science an instantiation of mode II knowledge production?
- who will be included in, and who will be excluded from cyberscience?

The session will not try to present one dominant theoretical or empirical framework. We will rather engage in provoking each other's curiosity in exploring this set of new phenomena and in critically analyzing it. Papers that serve this goal best will get priority treatment.

I reproduce this message in full in order to demonstrate the range of issues captured for Wouters under the term cyberscience. Wouters makes clear that the conjunction of ICTs and scientific practices raises a wide swathe of questions not easily subsumed under a policy interest in efficiency, collaboration, and large-scale science. The questions he raises are deliberately framed as provisional directions for exploration, based on extrapolation from existing preoccupations in the sociology of science and technology.

The qualities of cyberscience and its implications have indeed turned out to be issues of interest for STS. Subsequent to the session organized by Wouters in 1996, Nentwich (1999, 2001a, 2003, 2006) explored a range of questions in considerable depth from a technology assessment perspective, noting the extent to which change might be expected in various aspects of scientific communicative practice. Beaulieu (2001: 636) suggests that cyberscience may prove to represent "a novel, technologically-supported organization of knowledge-production, in which the digital format and electronic transmission of data figure prominently." She makes clear that the nature and implications of this novelty are emergent from practice, and as such are subjects for research, rather than issues to be extrapolated from the nature of the technology. In a recent collection of papers intended to display the state of the engagement of STS with e-science, authors note that dramatic wholesale change might not be apparent (Barjak 2006; Caldas 2006; Nentwich 2006; Palackal et al. 2006), and that there may be considerable specificity to the practices of different disciplines (Elvebakk 2006; Fry 2006; Haythornthwaite et al. 2006; Merz 2006). They also explore assumptions about the nature of science and the qualities of technologies that animate the current interest in e-science (Hine 2006a; Vann and Bowker 2006; Woolgar and Coopmans 2006; Wouters and Beaulieu 2006). Cyberscience provides a stimulating site for exploring the dynamics of the scientific endeavor and as such merits a closer examination, albeit one that remains agnostic about the absolute nature of change. When we take this approach, assumptions about the fast, cheap, and efficient nature of electronic communications, together with beliefs about the important aspects of existing scientific practices, may turn out to be more fragile than the grand visions suggest.

There has been a history of high-profile clashes between sociologists of science and some scientists who consider them poorly qualified to comment on science, or misguided in their approach. In the specific case of ICTs for science there may be potential for a more constructive engagement. The process of developing and promoting new infrastructures for science has led many involved to reflect on the circumstances that lead to successful innovation, and to acknowledge that social factors are key in determining whether a new infrastructure can be introduced on a sustainable basis. It has become relatively routine, then, among the growing numbers of practitioners of ICTs for science, to acknowledge that a key issue is understanding the social conditions under which new infrastructures will be deployed (Hey 2006). We might differ in what we mean by "social," but we have a meeting ground in that we consider qualities of the scientific endeavor that are often portrayed as trivial to be instead fundamental in the success of new infrastructures. As Wynne (1996) suggests, we can offer a sociological perspective as a means to encourage institutional reflexivity among practitioners and policy makers and to open up discussion about the contingencies of processes that might otherwise seem inevitable or closed off from debate. These are times of change, and of consciousness of change, and under these conditions there may be an opening for some productive rapprochement between science practice, science policy, and sociology of science.

In the advocation and use of ICTs within science on an unprecedented scale, we have therefore a phenomenon that merits the attention of policy makers, of practicicing scientists, and of sociologists of science, and that should help to provoke conversations between these domains. I will be using cyberscience to demarcate ICTs as a notable and remarkable feature of scientific practice, which have become thoroughly embedded in contemporary efforts to imagine how science should be in future. Without presuming that ICTs do indeed promote radical change in scientific organization and practice, this usage of cyberscience indicates an issue that deserves critical attention. As described in the introduction, the aim of this book is not, in the end, to arrive at an assessment of whether there has been or will be a radical enough change in the doing of science to deserve a distinct title. As Webster (1995) showed in his discussion of definitions of the Information Society, such demarcations are doomed to be infinitely open to dispute. Instead, the idea is to use the term to focus on a significant social phenomenon, and to consider in detail its manifestations at the level of policy and practice (which are, of course, not entirely separable domains). This focus stems from perspectives developed in the sociology of science and technology, and it is to these literatures that I turn in the next three sections to develop the approach in more depth, beginning with the ways in which the sociology of technology has theorized the relationship between technological innovation and organizational change.

Technologies and Organizational Change

Within the literature on sociology of technology there is considerable material to support skepticism on the potential of technologies to change a field of activity like science. Indeed, proposing a technology as an independent agent of change goes against the fundamental principles of constructivist sociology of technology as laid out by Mackenzie (1996), which recommend that we should:

- study technological development and its implications without subscribing to a determinist view that sees it as driven by an internal force of logic that then acts on society as an external agent;
- explain successful technologies without accounting for success by the superiority of the technology;
- take notice of beliefs about technologies and the trajectories of development, as these can drive innovation and prove to be self-fulfilling;
- realize that technologies which are adopted early on can acquire an advantage that leads them to succeed over others which might, in some lights, be judged better; and
- understand that relationships between technologies and society are mediated by knowledge, and take account of how that knowledge is distributed, shared, and contested.

Mackenzie's (1996) distillation of the constructivist perspective focuses on studying technological developments as they emerge without assuming that qualities of the technology are responsible for the trajectory that the innovation takes. Claims that are made about technologies and expectations about their impact are, however, significant factors within the complex of social dynamics that the constructivist sociologist studies. Viewing the problem in this way turns policy expectations about the potential of ICTs for science into a part of the phenomenon of interest, as much as they are competing views that the sociological account attempts to question. From this starting point we can set about exploring what computing means for contemporary science, without looking for a straightforwardly recognizable and singular "impact." It is salutary to look back at Attewell and Rule's (1984) exposition on why questions like "what do computers do to organizations?" are unlikely to find straightforward answers, and consider that asking what computers do to science is likely to be similarly futile. We will instead be exploring a complex array of social dynamics that promote and interpret technological innovations.

Science is not, of course, an organization in the sense that we usually use the term as meaning a social structure aimed at achieving a particular purpose. Science has some of the recognizable qualities of organizations, but it is more complex, and more diverse, than the term usually implies. If a laboratory is not instantly recognizable as an organization (Hine 2001), then an entire scientific field is even less so. Nonetheless, in the literature on technological innovation in organizations there are numerous elements that prove thought provoking for the distinctive case of science. The literature is rich in compelling yet contradictory case studies on how computers affect organizations. Here I will select a few authors who have addressed computers and organizational change as a complex process, and in particular look at discussions of the role of computers not in change in an absolute sense, but in talk about change. E-science, as described earlier in this chapter, has been a potent site for practitioners and policy makers to think about changing science. It therefore makes sense to consider perspectives from the literature on information technology in organizations that might illuminate this phenomenon. First, I will begin by using the work of Yates on the history of communication and information technologies in U.S. companies to think about the social processes that facilitate the entry of new technologies into organizations, and use this to reflect on the dynamics involved in technological innovation for science.

Yates (1994) provides a generalized model of the factors that influence the uptake and use of new information techniques and technologies, developed in the context of her work on the history of information processing and use in U.S. firms of the nineteenth and twentieth centuries. The model proposes three factors to explain variation in uptake: the size and structure of firms, which created a sense of the need for information; the supply of technologies and techniques for dealing with information; and managerial ideology, specifically belief in systematic management as a means to achieve efficiency. This latter ideology both promoted use of information-handling technologies and was in turn reinforced by their increased use. The upshot is that technological innovations and choices could be viewed neither as rational economic decisions, nor as created by the mere availability of technologies. Rather, technologies might be adopted for symbolic purposes as much as functional or economic reasons. Other organizations making use of particular technologies could act as role models whose practices were adopted regardless of more specific rationales. Although science is, of course, very different from the kind of organization that Yates discusses, her model provides some provocations for thinking about ICTs in science, and particularly the way in which new technologies might carry symbolism for practicing scientists and the institutions that employ them and for policy makers. Yates's model also supports speculation that different disciplines might adopt technologies in contrasting ways, while some disciplines might act as role models for others which they aspire to emulate in status.

Yates is against technologically determinist accounts that suggest that developments in technology alone were responsible for changing organizational culture and methods of control in organizations. The availability of technology can constrain or enable particular courses of action, but it can in turn also respond to developing concerns. Technology supported specific developments in organizational communication and control, but at the same time the direction of technological development was affected by the developing business priorities. Combining the two, new genres of communication emerged, facilitating the upward, downward, and lateral communication that was now perceived as vital for effective management. This observation chimes with Mackenzie's (1996) comments on the role of pronouncements about technologies as self-fulfilling prophecies. It also, as the next section will discuss, links with work in the sociology of science on the co-construction of scientific goals and the tools available for scientific work. Rather than influencing directions of work in a straightforward sense, new technologies can participate in the formation of new specifications of desirable goals, in turn influencing the way that technologies are perceived and the energies that are put into developing them.

Turning to the rate at which new technologies are adopted, Yates (1989) provides three case studies which show that the pace and the form of developments is highly variable. It was by no means inevitable even that growing and geographically dispersed organizations adopted new methods of systematic management involving control by communication, although their adoption was more likely in these organizations. Neither was the availability of technology a determining factor. In particular, she stresses the influence that key managers can have as powerful advocates of new methods: "The technology alone was not enough—the vision to use it in new ways was needed as well" (Yates 1989: 275). A similar point is made by Coopersmith (2001) in a very different context: the use of fax communications by political campaigning organizations in Texas. Coopersmith argues that "technologies do not diffuse automatically: people promote, adopt, and adapt them" (Coopersmith 2001: 64). Orlikowski et al. (1995) take this point a stage further, arguing that not just which technologies an organization adopts, but the ways in which that technology is used, can vary widely thanks to processes of "technology-use mediation" through which sanctioned advocates within an organization act as leaders in introducing technologies and molding the ways in which they are used. This makes clear that when looking at the development of cyberscience, it is important to be sensitive to the processes by which new technologies are made available, advocated, and interpreted, the locations in which this work is done, and the individuals and institutions by whom it is done. ICTs may not automatically be the solution for all scientific disciplines, and each discipline may make use of technologies in different ways, both because their culture and understanding of the problems to be addressed differ and because the technologies will be championed and interpreted in different ways. Addressing these issues obviously calls for a fairly detailed case study approach. In the next sections of this chapter I will discuss in more depth how we might go about exploring the ways in which new technologies are adopted in a specific scientific context. First, though, there is more to be said about the symbolic qualities of technologies and their role in beliefs about change.

In discussing the work of Yates (1989, 1994) I alluded to the assertion that technologies carry a symbolic quality. An organization might adopt a new technology not so much for rational efficiency gains as for the values or expectations associated with that technology. On this topic the work of Kling and colleagues from the field of social informatics is useful in capturing the extent to which beliefs about computing, as much as direct qualities of computers themselves, have provided a focus for organizational implementations. Iacono and Kling (2001) discuss the formation of "computerization movements," defined as the large-scale mobilization of support for computerization across contexts, formulating a package of expectations and values. This works, they suggest, on three levels. First, technological action frames make available a core set of ideas about what the technology is and what future it might bring about. Public discourses then circulate understandings of the technology, making them available to a broad audience through government, media, and diverse organizations and professions. A particular technological action frame shapes the public discourse, which is in turn appropriated at the local level of particular groups and organizations. Of course, the public discourse does not determine what happens at the local level, and there may be clashes between local experience and what the public discourse leads people to expect will happen. Computerization movements, just like other social movements, are heterogeneous, and they fluctuate in their emphasis and importance over time. The idea of the computerization movement provides a mechanism to account for the widespread use of particular new technologies and the diffusion of similar sets of expectations about what they will do, while still allowing room for diversity.

The concept of the computerization movement provides a provocative and useful way to consider how notions such as e-science and cyberinfrastructure are proving appealing across broad and varied constituencies. For example, the technological action frame of e-science is building sets of beliefs about what new technologies might do for science and mobilizing support across scientific disciplines and national funding regimes. These discourses are appropriated and interpreted at these various local levels, and although individual projects and experiences might diverge from the grand vision, the vision itself continues to disseminate. In the context of cyberscience, this is a reminder of the importance of broader societal beliefs about what ICTs are good for in shaping what scientists do with these technologies. Methodologically, it implies that understanding the development of cyberscience requires attention at different levels, including both observation of microlevel practice in the implementation of particular projects, and analysis of policy-level debates and documentation that help to articulate visions of what the technology is for. Without determining outcomes at the level of day-to-day scientific practice, policy-level input helps to shape the solutions that will make sense at the microlevel and offers resources for understanding goals and directions.

Returning to the question of whether and how technologies change organizations, Yates and van Maanen (2001) suggest that change is rarely experienced as radical. New practices may take considerable time to become established around a technology and be recognizable only in retrospect as "consequences" of technology introduction. Orlikowski (2001) writes of "emergent change" that is neither explicitly planned nor intended, but which can nonetheless be recognized in retrospect through analysis and/or by organizational members themselves. Orlikowski views technology not as an independent agent of change, but as a facet of organizational life that people adapt to, experiment with, work around and innovate in the face of. Simply by trying to get on with their work as they understand it, people can set in train what others might see as major changes through their local adaptations. The focus is on improvization as a feature of everyday life, leading to situated change. This is not to suggest that planned change is never experienced as such, or that technologies are never felt to be involved in organizational transformation. Rather, a variety of forms of change are to be expected, of which emergent or situated change is but one. What emerges from the adoption of a technology and the adaptations that go on around it is a "technology in practice" unique to each setting (Orlikowski 2000).

This perspective is stimulating for thinking about the ways that new technologies may be experienced in the context of scientific practice. Rather than expecting once-and-for-all statements about whether science has changed in the face of technological innovation we might expect to find both diverse practices and diverse opinions about those practices. In particular, expressions of change will be situated and will often not focus on those dimensions of practice that were predicted as the foci of change. New technology can act as a cultural disruption (Makagon 2000), providing the occasion for reflection on change, and articulation of aspects of culture. There will also often be a significant experience of continuity, as new ways of doing things complement, sit alongside, adapt to, and even stimulate the old ways, rather than replacing them in a radical transformation (Woolgar 2002a). Denying the possibility of radical change is certainly not to render grander visions, and in particular policy-level interventions, as irrelevant. As Mackenzie suggests, technological visions can become self-fulfilling prophecies, and as Iacono and Kling's (2001) computerization movements model discusses, technological action frames can offer a way of interpreting activities across diverse local contexts. Additionally, policy statements are not made in a vacuum. They have their own social dynamics and their own politics and cannot be taken as a direct portrayal of "what policy makers think." What is said at the policy level is influential for practice, but it is far from a direct prescription for action.

By extending interest from the grand-level pronouncements through to the daily experiences of work in science, we should gain a perspective on how hopes, expectations, and experiences of social change are organized around these technologies in a complex array of interlinked social processes. Looked at in this way, cyberscience becomes a manifestation of scientific culture articulated in the face of a new technology. Looking at how science acts in the face of new technologies and makes them its own tells us as much about science as it does about the technologies. It makes less and less sense, from this angle, to ask what the "impact" of ICTs on science might be. Rather, the project becomes an ethnographic one on a grand scale, to ask what the cultures of contemporary science are that are being enacted in, through, and around the use (or indeed, non-use) of these technologies. If we want to understand what contemporary science is like, exploring how new technologies like these are adopted and used is one powerful route into the problem.

The Co-construction of Science and Its Technologies

In the previous section I noted that science is not quite like the entities we usually describe as organizations. There is a need to be cautious in demarcating science automatically from other spheres of social life (Gieryn 1999), and there is a lot to be learned from the literature on ICTs in other domains. However, it is sensible to turn to the literature that focuses specifically on sociology of science to develop further the question of how technologies intertwine with scientific culture and practice, especially since the unusual role of technology in science has been one of the most remarked-on aspects in this field. Science is notable for the ways in which the view of the world it develops is mediated by various technologies that make aspects of the world visible and measurable. These scientific instruments are a basic component of the material culture of the scientific laboratory and of scientific fieldwork, and ethnographies of scientific practice have focused in depth on the way scientists work with instruments to develop factual statements about the world (Knorr-Cetina 1981, 1999; Lynch 1985; Latour and Woolgar 1986; Traweek 1988). Working with scientific instruments involved scientists in sorting out usable information from unwanted artifacts, enabling them to portray these instruments as ways to observe natural phenomena. Literary conventions dictated the ways that these results were then transformed and reported upon. As a form of practice and a field of culture (Pickering 1992), science was suffused with technological apparatus. ICTs offer to intervene significantly in this aspect of science, both as instruments in their own right and by mediating access to instruments and their outputs.

It has been suggested that the role of instruments in science has tended to be neglected at the expense of extensive consideration of the role of theory in representing knowledge (Baird and Faust 1990; Blume 1992; Baird 1993). Nonetheless, there is now a rich literature exploring sociological aspects of scientific instruments and their historical development. As they have developed historically, scientific instruments have become a distinctive class of technology (Warner 1990) concerned with allowing scientists to measure, visualize, and manipulate phenomena. The instruments that are available appear to play a large part in shaping the kind of science that results: as van Helden and Hankins put it, "because instruments determine what can be done, they also determine to some extent what can be thought" (van Helden and Hankins 1994: 4). New instrumentation can give birth to new directions for research and indeed to new disciplines, whether by borrowing from other fields, evolving an existing technique, or developing a completely new one (Gokalp 1990). However, the situation is considerably more dynamic than a straightforward determination of directions by technologies. Sociologists and historians of science have therefore looked at how the development and use of instruments help to define different areas of science and different sets of expertise, and how a new instrument becomes established as a reliable tool, once again turning the question from one of impact to a focus on process.

The development of new technologies for research often involves collaborations between different groups, each bringing a particular set of expertise but also each having their own identities, preconceptions, expectations, and working practices. This can produce "trading zones" in which these separate groups find ways to communicate and cooperate toward a common goal while still retaining distinct identities (Galison 1997). The development of instrumentation can often involve such cross-disciplinary interactions (Bud and Cozzens 1992) and also frequently involves commercial interests, although the likelihood of commercial involvement varies depending on the type of instrument in question (Riggs and Vonhippel 1994). Some instruments are developed primarily as commercial ventures, while others are a more grass roots–level innovation developed by scientists to serve a particular research goal. Skill, and its uneven social distribution, is a key issue in understanding how scientific instruments are developed and deployed. Collins (1975, 1985) showed that replication of an experiment could be a highly skilled technical achievement, albeit ultimately always open to doubt as to whether replication had been achieved successfully or not. Nutch (1996) describes how certain scientists may be able to distinguish themselves through developing specific skills and a distinctive affinity with instruments. Instruments are therefore implicated in scientific identity on both a group level and an individual basis.

Beyond special-purpose instruments developed for one-off experiments there is a category of what Joerges and Shinn (2000, 2001; Shinn and Joerges 2002) call "research technology": general-purpose pieces of machinery such as centrifuges, applicable to work in many different areas of science and engineering. Here the need for close collaboration and specialist skills is diminished, but processes of disembedding and reembedding take place, in order for the technology first to be detached from the particular circumstances in which it was developed and subsequently inserted into new contexts. Both development and use remain skilled domains, involving varying divisions of labor and flexible approaches to institutional boundaries, enabling what Joerges and Shinn call a "practice-based universality" (Joerges and Shinn 2000: 15) to develop. Importing an instrument involves adopting sets of vocabularies, values, and practices, both explicitly and implicitly. The spread of general-purpose instruments therefore becomes a form of communication between groups who might otherwise conceive of themselves as wholly separate and who can influence practice in subtle but significant ways.

These sociological and historical perspectives develop our understanding of the technologies involved in research by focusing attention on the different social groups where the skills to develop and to make use of scientific instruments are manifested. We are encouraged to disaggregate and focus on who is developing which technologies for whom, how they are enabled to work together, and how technologies mediate relationships in diverse ways. This provides a provocative agenda for studying cyberscience. The new ICTs may contain components both of general research technologies and more specific instruments. Some ICT applications are generic, whereas other are the outcome of specific projects aimed at producing a special-purpose tool. They may be developed and used by specifically skilled groups of people in particular institutional and disciplinary locations. It will therefore be important in what follows to explore who is developing and promoting these technologies, in what specific forms, and to what extent this has an impact on the social landscape of the discipline that is the focus of the case study. This area is particularly addressed in chapter 6, which explores the individuals, institutions, and initiatives that make up the emerging infrastructure.

There is another important dimension to innovative scientific instruments which focuses on their symbolic qualities and institutional location. It is useful to note that innovation can be valued for its own sake in science, echoing the observations of Yates (1994) on the symbolic qualities of IT innovation in organizations. Blume (1992) suggests that there comes a point when new instrumentation may be acquired by scientific institutions as much for reasons of status as for its specific functions: it becomes, quite simply, a matter of shame not to have the current piece of essential machinery. It is certainly worth considering institutional responses to cyberscience in this light. Chapter 3 makes a start on this project, by surveying the political landscape that developments in ICTs for systematics inhabit. The case study of systematics cumulatively demonstrates that ICT initiatives become something that systematics institutions cannot afford not to do, in terms of various policy-level pressures and interinstitutional rivalries, and as a response to broader expectations about what technologies modern publicly funded institutions should deploy.

In addition to the group dynamics and the negotiation of skill and identity that the topic of instrumentation raises, there is also the potential to consider more directly the role that instruments play in knowledge construction. The construction of a new machine often involves intensive labor both to get the machine to work and to convince relevant audiences that it is a reliable witness to the natural phenomena it claims to represent (Schaffer 1989) or that it is a working replica of the piece of experimental machinery it is supposed to reproduce (Collins 1985). This all occurs, it should be remembered, in contexts where there may be considerable doubt about the nature or existence of the phenomenon being measured, as well as whether the machine is functioning properly in measuring it. Subsequently, however, the instrument may become "black-boxed," its functions being largely accepted without question (Latour and Woolgar 1986; Latour 1987): a circumstance that can be viewed with some regret by established scientists (Blume 1992). All traces of the work that went into establishing a system of measurement and instrumentation can be, for practical purposes, forgotten (Schaffer 1992). The stabilization of a particular technology as part of the working practices of a discipline can change the skill base and disciplinary relationships, and can also more fundamentally affect research directions and the standards that are applied to judge useful and effective contributions.

Once established, a new technology can provide particular ways of conceptualizing problems and thus orient research directions. Computer simulation, for example, has become a form of practical methodology, allowing scientists to work with theories in experimental fashion (Dowling 1999). To the extent that they become accepted as offering worthy representations of objects, computers can provide a simulated laboratory (Knorr-Cetina 1999). The notion of artificial life has provided developmental biologists with new concepts to think with and new ways of exploring potential developmental pathways and influences (Keller 2002). Looking at other cyberscience applications that present data in digital forms, there is purchase in looking for the processes through which these new representations become accepted components of working practices within the discipline, and in turn are then able to facilitate the development of new practices and theoretical directions.

In relation to cyberscience, Beaulieu (2001) argues that the introduction of digital resources to store and pool brain images has contributed to the development of new standards of objectivity in the field. Similarly, Daston and Galison (1992) argue that the arrival of photography and other forms of mechanical reproduction prompted a new articulation of objectivity standards in science over a century ago. It is important, then, in relation to digital tools as much as other research-related technologies, to attend to the processes by which technologies become accepted as valuable additions to the research environment and to remain alert to the possibility of unanticipated consequences for scientific knowledge and practice. In none of these cases, however, is it enough to say that the new technology changes science. As we saw in the previous section, new technologies are developed and deployed within institutional contexts, and their consequences can vary dramatically in relation to changing social dynamics of use.

In the context of science, previous work has warned us to expect the technologies of scientific research to be made meaningful only as a part of the assemblages of physical space, working practices, and material artifacts that Lynch (1991) calls the topical contextures of science. Technologies are not transferred naked across social boundaries, but travel packaged with theories and practices to make them meaningful (Fujimura 1987, 1988, 1996), and this is as true of digital resources as it is of material tools and techniques (Fujimura and Fortun 1996). The process is one of co-constructing the tools for the work and the nature of the task itself, developing over time standardized and stable ways of working that fit in with scientists' understandings of the prevailing funding regimes and policy context (Clarke and Fujimura 1992). We need to ask, then, how particular ways of working with new digital tools arise and become stabilized, and how this process interacts with the definition of classifications, tasks, and priorities (Bowker and Star 1999). New technologies, seen from this perspective, participate in the shaping of disciplinary directions and outcomes, but as a component of a dynamic social field rather than externally originating forces that affect otherwise static domains.

In studying the dynamic social processes that produce and legitimize technologies in scientific research it is important also to recognize that the way scientists discuss these technologies varies considerably with context. I have previously noted that, in the case of information technologies in genetics, we can see varying narrative representations (Hine 1995). I argued that information technology as a scientific instrument emerged from a complex and varying co-construction of the scientist as an actor with the instrument and the genetic object. Sometimes scientists talked about problems with the instrument, about their own fears and skills, and about impressive or worrying developments. At other times the instrument receded into the background, and scientists talked of their own agency in looking at candidates and pulling out genes, or spoke of genes as independent phenomena, capable of announcing themselves unaided. As the machine takes a smaller part in the narrative, more space is made for the emergence of skilled scientists and real-world phenomena like genes. These varying narratives suggest that the role of computers in research is an important rhetorical phenomenon as much as a component of practical scientific work. Thus it will be useful in studying a cyberscience to develop an ethnographic sensitivity to what it means to scientists, to institutions, and to policy makers to talk about involvement in ICTs, and to look at the ways in which talk about technologies is implicated in the construction of identities for scientists and institutions and the attribution of qualities to scientific knowledge. Within biology in particular, computers have been a potent source of visions for a transformed discipline (Lenoir 1998b; Hagen 2001, 2003), and it is crucial to keep in mind that such technology-talk is a powerful form of rhetoric.

This brief review of a small portion of the considerable literature about the social dynamics of scientific instruments, in production and in use, provides many leads to pursue in understanding the development of cyberscience. The notion of co-construction offers a very useful way of capturing these dynamics. Rather than having an impact on science, new technologies can be thought of as co-constructed with the research environment. What the technology is, who the scientists involved are, and what they aspire to achieve are co-produced. Scientists may not have a fully formed idea of the direction they want to pursue in advance of the technologies' being available to achieve the task: instead, processes of accommodation and imagination bring goals, technologies, and scientists into alignment. In this regard science is not independent of wider cultural trends or of prevailing beliefs about what is to be expected from technologies. Studying cyberscience involves looking at the development of technologies that sometimes originate within science, sometimes outside it, but are in many cases also connected with other cultural spheres. The incorporation of this particular set of new tools into scientific practice will involve working with these other beliefs and expectations about the technologies.

One particular topic on which there has been considerable speculation across a wide array of cultural domains is computer-mediated communication. In the next section I will therefore focus more closely on this area of technological innovation, beginning with a discussion of the communication practices of science before broadening out to a perspective on how new scientific communication practices become embedded and acquire social meaning.

Constructing Science through Communication

Thus far I have discussed technologies in the form of scientific instruments, entwined with the skilled work of knowledge production. This provides some provocative ways for thinking about new ICTs, but it does not directly address one of the most obvious places where they are likely to be involved in new scientific practices: communication processes. In this section, therefore, I discuss literature on the importance of communication in scientific practice, the various forms it takes, and some suggestions about the likely impact that new media could have on knowledge production. Along the way this brings into question some of the expectations around communication in science introduced in the section on policy interventions, and also casts a new light on the processes of co-construction and imagining already identified as important for scientific instruments.

The importance of communication to the scientific endeavor depends in part on the dual nature of science as both localized practice and universal knowledge. One important aspect of scientific communication involves the passage of knowledge from the private territory of the laboratory into the public domain, making it available both for consumption by other scientists in their ongoing work and for application in technology, medicine, engineering, and the like. The formal scientific communication process is about taking the products of local practice and making them universal. As Latour and Woolgar (1986) describe it, fact construction is a process of progressively stripping away the modalities that qualify and localize claims to knowledge, involving a cascade of inscriptions, progressively more abstracted from the local circumstances of production (Latour 1987). Knorr-Cetina (1981) shows how, in practice, the writing of a scientific paper is a process of fine-crafting the presentation. A scientific paper, then, is a rhetorical achievement which subsequently gives the impression of being a straightforward reporting of what was done (Myers 1985). Or, to put it another way, scientists have learned to use the technology of the journal article in a conventionalized way, which involves presenting a version of scientific practice that strips out local contingencies. It is this kind of communication practice, as much as the physical isolation of the laboratory, that maintains the notion of science as a special form of knowledge independent of social influences. Nevertheless science is a field of cultural production (Lenoir 1998a), and sociology of science has demonstrated that if we want to understand that field it is important to study the emergent communication practices which underpin and enact that culture.

Formal scientific communication appears to be challenged by new media, particularly since they offer the possibility to transform existing institutional structures for publishing. The prospect of bypassing journals as the outlet of choice for scientific publishing by publishing on the Internet, whether on an individual basis, in online journals or preprint repositories, or in publicly supported archives, brings into question many aspects of the previously accepted system. It is intriguing to speculate whether new communication practices might entail the revision of the knowledge production process itself or the forging of new relations between science and its publics. If the fine-crafting of the scientific journal article is so important to the knowledge construction process (Knorr-Cetina 1981), then changes to its form may meet resistance, or prove not to be attractive or sustainable. It is notable that preprints are widely distributed in electronic form but still retain the same literary presentation as journal articles. Discussions about questions of quality control and intellectual property, the role of commercial interests, and the timeliness of publication have indeed all been raised in connection with online publishing (Nentwich 2003).

Rather than wholesale unidirectional change, it appears that the new technologies have offered a cultural disruption and prompted reflection on the established system. Whether or not calls for scientific research publications to become open access are implemented in the longer term, in the short term they have prompted unprecedented levels of debate on the function and operation of the scientific publishing system. Similarly, the advent of digital libraries has prompted widespread research, not just on digital libraries themselves but in a wider sense on the functions and institutional locations of libraries in general (Chowdhury and Chowdhury 1999; Borgman 2000). Where new ways of communicating scientific facts are envisaged it is therefore interesting to consider the discussions that result and to examine how the processes of scientific knowledge production adapt. It is interesting both to speculate on the possible outcomes of change and to study the dynamic processes through which change and continuity are negotiated in practice.

The potential implications for knowledge construction processes are not confined to reporting of results in the formal literature as described above, but also extend into working practices at all stages and particularly may affect informal communications. In the formal journal literature many of the matters of skill that are needed to get laboratory equipment to work are stripped away from the published account, and scientists acknowledge that the literature is not the place to go in order to find out how to get a particular technique to work (Lynch and Jordan 1995; Jordan and Lynch 1998). For this, scientists habitually rely on a range of less formal communication methods, including face-to-face contact within the laboratory or at conferences, visits and exchanges between laboratories, and recently the informal opportunities provided by online forums. Electronic communications provide the possibility of extended opportunities for informal communication, enabling scientists to consult one another across institutions and form interest groups where this kind of information can be shared on a casual basis. Again, this could be a matter of changing working practices on an everyday basis, but it might also embody a shift toward new structures that cut across institutions, transforming the organization of scientific work on a more gross level. The association of electronic forums with informal communication may be yet another instance of co-construction. The nature of computer-mediated communication is constructed in and through the things that people find sensible to do with it, and this is not determined by particular qualities of the medium itself (Hine 2000). It is not inevitable that online scientific forums will be used for informal communication, but it is certainly possible for them to develop in ways that serve this function as long as they share enough common ground for specialized language and technical issues to be mutually understood (Hine 2002).

Hert (2003) describes a scientific discussion group that he observed as developing a "quasi orality," containing aspects of other communication genres in science such as lecture notes and conference papers, but also having significant qualities of spoken interaction. This genre allows for a certain amount of mutual interrogation and clarification and for the emergence of some sense of community within the forum. Alongside their explicit functions, as announced in their formal remit, forums also offer implicit benefits and subtler connections, forged in the way that scientists collectively come to use them. Forums can foster group identity and help scientists to feel part of a collective effort. Hert (1997) found participants skillfully appropriating qualities of the medium, particularly the ability to quote selectively from previous contributors in order to take control of interactions. This serves as a significant reminder that communications media are available to be appropriated within specific disciplinary cultures and by motivated individuals. As Hert (1997) describes it, there is "mutual elaboration between the discussion and the medium used": ideas about what to do with this particular communications technology in this particular circumstance are worked out in the moment, as discussion goes along, and reinforced as other participants take them up and work with them. Considerable diversity is therefore to be expected in the ways that scientists use computer-mediated communication, including disciplinary differences (Walsh and Bayma 1996a,b; Kling and McKim 2000; Walsh et al. 2000; Matzat 2001; Barjak 2004), just as we find diversity in scientists' use of other information technologies (David and Zeitlyn 1996; Zeitlyn, David, and Bex 1999).

Lewenstein analyzes the use of communications media in the cold fusion debate to suggest that scientists use a variety of outlets to publicize and discuss their work, and that there are complex interactions between these different sources (Lewenstein 1995b). On the specific subject of online forums, he found that problems of filtering out extraneous information and of forming judgments about reliability hampered their usability as formal communication mechanisms for scientific knowledge, but that a certain amount of detailed information on practical issues was transmitted in this way (Lewenstein 1995a). The functions of online forums are, though, not confined to exchanging information about practical issues. Matzat (2004) found that another key benefit scientists experienced was in building up social capital through informal links that then helped to establish an audience for their formally published work. Matzat (2004) also reviews the early literature that suggested computer-mediated communication might have an equalizing effect, allowing previously marginalized scientists to become included in a wider scientific discourse and promoting collaborations, and finds this optimism largely misplaced. Davidson, Sooryamoorthy, and Shrum (2002) do find evidence that computer-mediated communication is changing ways of thinking about research for scientists in Kerala, but many problems and constraints get in the way of realizing their visions.

Although online forums have become a routine and valued part of communication across diverse areas of science, the effects are patchy and there is little sign that in this kind of forum all scientists are equal. Quite the reverse, there seem to be considerable possibilities for scientists to distinguish themselves and accrue social capital. Being visible in an online forum by adding to discussion and answering queries can be a way to become known in a disciplinary context. There is little sign, however, that this kind of venue is replacing the more traditional contexts of conference and coffee-room gossip (Nentwich 2003) and the usual trappings of the credibility cycle (Latour and Woolgar 1986). In the context of high energy physics, where as previously said collaborators are accustomed to working together while physically apart, it is notable that collaborations are still generally begun offline, and only then taken online, largely to restricted circulation email (Merz 1998). According to Merz, although public forums are available, they have not proved appealing for finding collaborators or for reproducing the kind of opportunistic informal chats that can arise in the public areas of laboratories. This kind of observation brings into question the policy reliance on ICTs to mediate collaboration, and suggests that we need a situated approach to understanding the ways that particular modes of communication make sense to specific groups of scientists.

In addition to group interactions provided for by mailing lists and newsgroups, electronic communications on a one-to-one basis have been very appealing for scientists. Email has become a part of the taken-for-granted communication practices of science. Walsh et al. (2000) showed that it was near universal among experimental biologists, mathematicians, physicists, and sociologists, and that its use was associated with higher levels of productivity and more collaborative work. Scientists frequently use email to coordinate long-distance work, overcoming some of the difficulties raised by working apart and answering questions, particularly where these relate to practical issues, with quick answers. Email retains considerable appeal as a means of having informal communication across distance with known collaborators, in contrast to the online groups where one addresses an unspecified multitude of colleagues. Personal Web pages have also become common among scientists and other academics, although again their use may well vary widely across different knowledge production cultures. Web sites and the hyperlinks between them can provide the material for exploring collaborative networks and other forms of interaction between scientists (Harries et al. 2004). Again, it is an empirical question to be answered on a very specific basis how particular disciplinary cultures adopt and adapt to email and the Web, and which aspects of their work they use them to support.

In addition to threats to existing models of publication and informal communication, new information and communication media have also offered altogether new models. In biology, databases have formed the basis for a new communication regime allowing genetic data to be stored and retrieved from centralized data repositories (Hilgartner 1995). The communication regime is defined by Hilgartner as an established network for the dissemination of science, comprising an actively constituted system of technical and social arrangements. Databases in themselves are not communication regimes, but they can provide a focus for such regimes to develop. Hilgartner and Brandt Rauf (1994) suggest that we need to study how data is packaged, made available for sharing, and used under such regimes, bearing in mind that access to resources occurs within complex research networks and can take place via diverse routes and mechanisms. Databases such as the large public repositories of genetic sequence data provide one highly significant means of sharing data for contemporary biologists, although there are contradictory pressures over what to share, when, and how (Hilgartner and Brandt Rauf 1994). Arriving at the current point where such data-sharing initiatives are routinely accepted (Brown 2003) involves dealing with such issues as establishing agreements with journal publishers and developing systems for allocating credit and managing intellectual property, in order to make sharing data a feasible and meaningful thing to do (McCain 1995, 2000; Hurd et al. 2002; Beaulieu 2003).

Continuing this line of thought about the interaction of new forms of data storage with scientific practices, Derrida's (1998) analysis of the archive provides some highly provocative thoughts to illuminate the relationship between a scientific discipline and its technologies of recording. In particular he provides a framework for thinking about databases as contextual objects, rather than as technologies with independent effects. There are some problems with taking Derrida's thinking as automatically applicable for our purposes. Derrida is writing in a very specific context; he speaks on the occasion of the inauguration of Freud's archive, and thus interweaves his thoughts on archiving with his readings of Freud, in order to illuminate the incongruities and resonances of the practice of archiving with the practice of psychoanalysis. Much of the work is thus taken up with exploring the issue of memory in Freudian thought. There are, however, some moments where Derrida chooses to generalize, and to say something about archiving that we can take to apply to data-sharing technologies outside of the specifically Freudian context.

First Derrida takes us on an exploration of the origins of the word "archive," in order to persuade us of certain qualities that an archive must have. The root of the word has two connotations: it suggests the archive as both a commencement and commandment, that is, a point of origin and a site of authority. In order for it to function, the archive must have a place, and a guardian. This "privileged topology" provides for a consignation, or gathering together of signs, such that the corpus becomes an ordered system. For Derrida, then, the archive is not guaranteed by placing items together. Rather, it has to be institutionalized, located, and ordered. This observation seems to have a general applicability to archiving outside of the Freudian context. Whatever the technology used to build the archive, and whatever the subject matter, Derrida's observations serve as a reminder to examine who is doing the archiving, in whose name, where, and under what claimed authority. In an aside, Derrida notes that this perspective raises a concern with the politics of the archive, and he suggests that we might measure democracy by how far there is participation in the archival process: a true democracy would allow open participation in the telling of history. Indeed, the claim to hold the authoritative version of a topic is highly politically charged. As Bowker (2000) describes, the practice of combining biodiversity databases and attempting either to create authoritative versions or to avoid doing so is highly delicate and problematic work.

Derrida, then, considers that an archive is a particular form of institution, and that as such we need to consider where it is located, how it is ordered, and by whom it is guarded. He does, however, also consider the role of the specific technology on which the resulting archive depends. In the particular context that concerns him in *Archive Fever*, he considers whether the Freudian model of the psychic apparatus, and hence the theoretical basis of psychoanalysis, would be any different if Freud's thinking had drawn more on archival technologies than a "mystic pad." This is an intriguing avenue of speculation: how might Freud have thought about memory if he had lived in the Internet age? Derrida seems to adopt a highly situated notion of psychoanalysis, as a form of knowledge for its time rather than a transcendent truth. We can also ask this question of other disciplines, wondering how the models that shape particular directions of inquiry are in turn shaped by particular cultural and technological surroundings. The technology available to the scientist might, for Derrida, affect the knowledge produced by providing the concepts to think with: as chapter 4 will show, databases have offered a potent symbol for herbarium and museum curators to use in thinking about their collections.

As well as the archival technology's role in providing models for memory, Derrida also suggests more direct effects on knowledge. Most pertinently for our purposes, Derrida asks if psychoanalysis would be any different if, instead of handwritten letters, more recent communications technologies had been available for storing the thoughts and actions of Freud and his followers. Derrida is sure that psychoanalysis would indeed be different, in some fundamental ways. He says: "No, the technical structure of the archiving archive also determines the structure of the archivable content even in its very coming into existence and in its relationship to the future. The archivisation produces as much as it records the event" (Derrida 1998: 17). Specifically, Derrida turns his thinking to the most prominent new communication technology at the time—email—and states that psychoanalysis would have been different if email, rather than handwritten notes, had been used. The notion of an archival technology defining what is to count as an event in the first place casts doubt on the usual idea of an archive as a true reflection of what went on. For Derrida, "what went on" is arrived at only in conjunction with the act of archiving that purports to record it. In the case of the Freudian archive, the handwritten records have come to count as what went on between Freud and his patients, correspondents, and followers. Had email been available, Derrida suggests that our understanding of those events would be quite different. We can ask similar questions of other scientific disciplines, and their systems of publishing and archiving, considering how the progress of a discipline and its forms of knowledge might have been different had other technologies been available. Of course, this has to be a speculative exercise, and it is easy to fall into a deterministic view of the science that particular technologies might have created. All the same, the exercise of thinking about scientific knowledge as shaped in particular technological and communicative milieux is a provocative one.

A further commentary on the role of the archive in contemporary academic life is provided by Myerson (1998). According to Myerson, Derrida's perspective on archive fever is but one of the possible, and coexisting, responses that the archive evokes. Whereas the Derridean archive is a source of fever, for others its danger is precisely that it is a cold medium, one that fixes and decontextualizes. Against archives, then, we have opposing dangers of fever and chill. Both invoke difference from previous forms. Myerson, however, also explores modes of thinking that would situate the electronic archive as a continuation of existing traditions of rationality and the urge to grasp and order things, or as a new way of performing rigorous analysis. Myerson views the electronic archive as a thoroughly ambiguous object, evoking responses that find it simultaneously alluring and dangerous, mundane and exotic. This analysis, then, is a final reminder not to attempt to close down too quickly on the implications of ICTs in science. Reactions are likely to be elusive, varied, and contradictory.

In the discussion so far I have tended to portray formal scientific publishing, informal communication, and databases as if they were all separate. As discussed in earlier sections of this chapter there have been moves, more recently, to institute online environments for doing science that would allow access to large collections of data, instruments, and tools for analysis, and which provide the full spectrum of communication mechanisms. Collaboratories offer a way of deploying ICTs to give a new form of organization for scientific work (Finholt and Olson 1997). There are considerable problems, however, in innovating on this scale. Many of the developments in computer-mediated communication for science that I have described so far have been initiated at a grassroots level, allowing for relatively low resourcing and an experimental approach to development. A newsgroup, a mailing list or even a preprint archive can be set up on a fairly small budget. A larger-scale infrastructural development, such as a data repository or a collaboratory, can require much greater input to develop a complex of tools and facilities and, thanks to its complexity, will depend on a much greater detail of prior specification of appropriate uses. As Star and Ruhleder (1996) describe, it can be challenging to sustain the level of resourcing needed and to manage the complexities produced by identifying and balancing conflicting user needs. A new form of publishing needs to gain recognition from scientific institutions, governments, and funding bodies if it is to become thoroughly embedded in scientific practice. Chapter 6 discusses the difficulties inherent for institutions in deciding which initiatives to invest their efforts in, and the practical politics involved in making initiatives happen.

New ICTs clearly offer a variety of ways in which science can be communicated, encompassing informal interactions, formal publishing, and the ability to make data available on a scale not previously possible. There is also a prospect that different communication forms could be combined into larger-scale research environments, providing an infrastructure for performing scientific collaboration at a distance. These new technologies are, however, not enough on their own to change science (Bierly 1988). In each case, technologies have to find a place within the cultural context of science, and their use has to become accepted as a part of routine scientific practice. The introduction of new communication regimes can involve the explicit specification of aspects of science otherwise left tacit, and can introduce tensions where established practices of sharing and ownership are brought into question. These new technologies do have the potential to occasion changes in scientific practice by providing new forums and venues for scientific communication, by allowing for different kinds of spatial and temporal organization in the knowledge production process, and by prompting reflection on existing communication structures; but achieving this outcome is the upshot of a complex set of dynamics. Simply by being recognized as new a technology can occasion cultural disruption and this has often been the case with computer technologies.

One fruitful way to think about the dynamics which accompany the introduction of new computer-mediated communication infrastructures in science is to return to the notion of co-construction. Just as new scientific instruments can participate in the co-construction of goals, identities, and technologies, so too can computer-mediated communication be thought of as a co-construction of the technology and the uses to which it is put. The process which Hert (1997) called mutual elaboration of medium and discussion can be seen as a co-construction of the goals of the participants in the discussion and the properties of the technology which provides for the discussion. The co-construction of technologies and users is a familiar theme from the sociology of technology (Oudshoorn and Pinch 2003). Similarly, the development of a new communication regime around a database can be seen as a co-construction of what it is that databases do, what the particular community using it wants to achieve, and what is culturally feasible to bring about in terms of structures of reward and accountability. Technological capacities emerge in tandem with the identities of users and developers and the actions they carry out using them.

The case study which forms the rest of this book is an opportunity to explore in depth some of these processes of co-construction and the cultural disruptions that they entail. Chapter 5 focuses specifically on communication, although the overall argument is that developments in communication are interwoven with aspects of the political, material, and institutional culture of systematics discussed in other chapters. The co-construction of ICTs and scientific practice forms one site where these aspects are brought together in the imagining of the contemporary scientific endeavor. In the next section I turn to discuss the methodology that underpinned the case study and allowed for this outcome. It will be apparent by now that the approach I adopt depends on ethnographic sensitivities. Nonetheless, studying cyberscience turned out also to be an opportunity for creative appropriation of the ethnographic approach, and therefore the methodology requires some more specification and justification than the hints that have emerged thus far.

Rationale for an Ethnographic Approach to Cyberscience

By considering literature that theorizes the situated relationship between technologies and organizational change, and by looking at research into the links between technological developments and scientific practices, I have focused on two concepts useful in the understanding of change: imagining and co-construction. Instead of looking for objectively identifiable change, there is purchase in looking for moments of imagining where ideas about change are articulated. This imagining is not a matter of a stable set of individuals portraying their past and their future. Rather, imagining is a moment of defining both the territory and the beings who inhabit it. The notion of co-construction provides a way to think about the emergent properties of the imagined and the imaginers. Co-construction brings into focus the sites of imagining in an emergent sense, where imagining a potential future inevitably involves specifying the present condition against which it is to be viewed. By articulating their moves into the future through the discussion of desirable technological developments, practitioners and policy makers in contemporary science form a commentary upon who they are and where they are now. Imagining also has a politics: it proves significant to explore who is imagining whom, and who the audience is for their imagining.

The conceptualization of technological innovation in terms of situated processes of imagining and co-construction provides an organizing rationale for the fieldwork that I carried out within systematics. By reviewing some key literature from history and sociology of science and technology, I have established an array of issues that a sociohistorical understanding of cyberscience could address. The issues of concern span:

- commentaries on the past, present, and future of ICTs within scientific work, to be viewed as situated practices that are produced by miscellaneous interested parties and aimed at diverse audiences;
- cultural expectations about ICTs as they connect with and are reshaped in dialogue with science policy and practice;

- the integration of ICTs with the knowledge production process, both as scientific instruments in their own right and as they mediate access to material objects and other instruments;
- the manifestation of ICTs within existing and novel discipline-specific communication regimes; and
- the institutionalization of ICT initiatives as played out in conjunction with diverse individual, institutional, and disciplinary identities.

Taking these concerns on board, cyberscience becomes a complex phenomenon inhabiting a range of sites and forms of expression and time scales. It is important both to be alert to the rhetorical and symbolic qualities of claims about ICTs and at the same time to acknowledge that these technologies are potent sites for change to be experienced and articulated. Taking this perspective suggested a range of potential fieldwork sites: going inside institutions on the level of daily practice, and also looking at the ways in which individuals, institutions, and the discipline were manifested more publicly in the various forms of communication within and about the discipline, or looking at the online venues where data was shared and discussions conducted. I sought to explore the constitution of the different groups involved and the ways in which innovation in ICTs was significant for their identity. I aimed also to explore the communication practices that resulted from the social dynamics of ICT production and use and to investigate the origins and implications of new communication contexts that emerged as new communications technologies were deployed. I sought to understand where the impetus for innovation came from, how it filtered through the discipline, and how it was experienced in terms of continuity and change.

It would be possible to focus on any one of the issues outlined above, but I have chosen to pursue all of them simultaneously, in order to explore their interlinkages and mutual elaborations. This problem area seemed to me to be appropriate for a multisited (Marcus 1995) approach, both because the phenomenon that I studied crosses geographic space as a matter of course and is embedded in multiple contexts, and because I had a hunch that there were multiple strands of meaning to pursue in order to work out how particular uses of ICTs made sense for practitioners. This multifaceted approach would, however, inevitably entail some loss in terms of depth of ethnographic attention to daily practice. A strong heritage of laboratory ethnographies (including notably Knorr-Cetina 1981; Zenzen and Restivo 1982; Collins 1985; Lynch 1985; Latour and Woolgar 1986; Traweek 1988) demonstrates the purchase of this intensive focus on practice (for more adequate introductions, see Pickering 1992; Lynch 1993; or more recently Sismondo 2004). In describing how scientists worked with machines and material objects, used instruments to provide traces of natural phenomena, identified usable information and discarded artifacts, transformed inscriptions and produced literary descriptions of their activities, these pioneers developed a view of science as a form of practice and a field of culture (Pickering 1992). Including within their scope how scientists came to be able to make claims about the world, together with how their social worlds were organized in terms of their careers, their institutions, their working practices, and their communication regimes, laboratory ethnographers provided a view of science that at the same time emphasized its specificity and showed how it could be considered as producing knowledge that was constitutively social rather than, as often portrayed, asocial.

The form of the laboratory ethnography has some resonance with the anthropological tradition of global ethnography (Fog Olwig and Hastrup 1997; Burawoy 2000), and has a considerable appeal as a way to get close to the lived experience of operating within the complexities of contemporary cyberscience. This is closer to the kind of approach that Beaulieu (2001, 2003) has taken to the understanding of new infrastructures for science, and it provides a very promising model for future work on emergent practices in cyberscience. In the current work, however, I explicitly set out to go in search of different sites, as an appropriate choice for the kind of intervention that I wished to make in understanding the diverse yet interconnected social worlds of cyberscience. I therefore focused my investigation provisionally at the level of the discipline as my unit of analysis rather than focusing on a particular institution or group of systematists. Multisited ethnographic approaches have been persuasively articulated by Marcus (1995) and, as Hess (2001) describes, this more wide-ranging version of ethnography has been particularly fruitfully adopted within STS. Heath and colleagues (Heath 1998; Heath et al. 1999) provide models for exploring how science is constituted by different interlinked locations. Martin (1998) shows how an anthropological approach to culture suggests that, instead of only spending time in a particular place, like a laboratory, ethnographers can also usefully pursue flows, interactions, and connections. These mobile and connective approaches to ethnography begin to realize the prescription from Latour not to assume that there is an inside and an outside to the laboratory (Latour 1983). By tracing and following connections between sites, both connections between laboratories and beyond, we can realize a mobile ethnography that aims at an understanding of the ways in which the contemporary locales of scientific practice are created, extended, and connected. The study becomes a mobile sociology of science (Urry 2000), taking into account the cultural complexity inherent in accepting that scientific culture is intimately connected in complicated ways with other cultural spheres (Hannerz 1992).

Ethnography, though compelling in the depth of understanding it gives of how diverse cultural elements are dynamically experienced and reproduced in real time, can be less suited to understanding change and stability and the coordination of scientific practices across time and space. It is not accidental that historical methods, rather than real-time ethnographic methods, have often been used to understand the transformations involved in organizing science across different sites and different social worlds. The historian has a kind of social mobility that is harder for the ethnographer to achieve: by taking a perspective distanced in time the historian also becomes socially mobile, looking at what was happening at the same time in different places. Keating, Cambrosio, and Mackenzie (1992) make a similar point in their history of the affinity/avidity debate in immunology. They suggest that ethnographies, focusing on local sites such as the laboratory, tend to underplay the importance of disciplinary frameworks in lending continuity to scientific practice. In their analysis of collaborative work in the development of antibody reagents, Cambrosio, Keating, and Mogoutov (2004) go further, arguing that neither ethnography in specific sites nor simple quantitative indicators adequately capture the dynamic and varied and complex phenomenon they aim at. They therefore combine interviews and analysis of documents with a computer-assisted network analysis. Without adopting a formalized approach to network analysis, I felt that the phenomenon I wished to characterize also lent itself to deployment of an imaginative conceptualization of its field site, and benefited from some supplementation of conventional fieldwork procedures with more structured analyses and visualizations, particularly in relation to an exploration of the Internet landscape.

The kind of connective ethnography used in this case study is closely linked to the form of virtual ethnography I have proposed previously (Hine 2000). A core tenet of this approach to ethnography of computer-mediated communication is to bear in mind that these technologies both form sites for the enactment of culture and are cultural artifacts that come to carry meanings for their users. In order to understand how what people do with computer-mediated communication makes sense to them, it follows that we need to look closely at what it is they do online as a cultural manifestation in its own right and also to ask what it is that makes their online activities meaningful, both in the other aspects of their lives and in relation to the online phenomena they encounter. Bringing this perspective to bear on cyberscience provides a provocative way to tie together the various sites of interest that I have identified, by formulating the goal as a connective ethnography of the way that use of ICTs makes sense within my chosen field of contemporary science. As cultural sites ICTs have the potential to become key ways in which contemporary science is enacted, and it is therefore important to explore how these sites of science are constituted. At the same time, ICTs are cultural artifacts which it makes sense for scientists to use thanks to diverse aspects of the broader disciplinary culture that ICTs both inhabit and enact. In the spirit of Silver's (2000) call for critical cyberculture studies, a connective ethnography of cyberscience could give close and detailed attention to what these people do online, to the stories they tell about these kinds of activity, and to the social, cultural, political, and economic issues and the decision-making processes that frame what these technologies are to them. Broadening out from the Internet, as in my previous attempt at virtual ethnography, to ICTs in this case acknowledges the interconnected legacy of various forms of data storage and analysis that precede and accompany use of computer-mediated communication. For cyberscience it makes little sense to start with a narrow focus on computer-mediated communication as an isolated phenomenon.

The field site in a study such as this is not a predetermined entity. The connective approach depends on beginning from common understandings about where the phenomenon being investigated happens and what other issues it is linked with, using the ethnographic sensitivity progressively to identify other sites and connections and to question the ones identified at the outset (de Laet and Mol 2000). As has been discussed within anthropology in recent years, the field is a construct of the ethnographer (Gupta and Ferguson 1997; Amit 2000), and as such we are responsible for building appropriate field sites for the exploration of issues that concern us (Marcus 1995, 1998). The ethnographer's role is to achieve a sensitivity to the way that meaning-making is structured, which may depend on spatially complex and unexpected forms of connection. This kind of approach calls for ongoing reflexive shaping of the project (Reid-Henry 2003), using any media that become appropriate, as Sunderland (1999) suggests, and using this engagement to reflect on the ways that social life is constituted through their use (Hine 2000). Online and offline connections are pursued back and forth as they promise mutually to contextualize and offer meaning for one another (Leander and McKim 2003; Orgad 2005). Taking this approach to my study of systematics offers the important dimension of learning about the texture of social life in contemporary systematics through its deployment of a variety of communications media. In echo of my previous remarks about virtual ethnography (Hine 2000), the ethnography of cyberscience is conducted in, of, and through the medium of ICTs. The Internet can be thought of as a mirror of systematics and as such is a fruitful ethnographic field site, but we need to move around and beyond online observations if we are to find out what kind of mirror it is for its users: whether it is a fairground distortion, a true reflection, or a flattering artifact in a fairytale.

By pursuing connections that seem to make sense in terms of understanding the practices and preoccupations of a particular phenomenon, this form of ethnography produces a result that claims to be sensitive to portraying social life as lived without being an objective or total depiction. Connections are simply too multiple and too diverse for any ethnographer to be able to claim to have covered a whole territory or to have found out in any absolute sense what constitutes a particular phenomenon. In its reliance on experiential knowledge and the body of the ethnographer as the instrument of research, any form of ethnography is inherently partial, being limited in the extent to which other forms of life can be experienced while also living one's own. The connective approach to ethnography increases consciousness of this partiality, since there is always awareness of some connections missed out, some opportunities ignored, and an appreciation of the untraceable multiplicity of connections forged in the moment that help lives to make sense. Despite the aim to remain open to any kind of connection that might suffuse the experience of systematics with ICTs and help to make this experience meaningful, there is therefore nothing total about the account contained in the following chapters. It is a presentation of a selected set of issues that together help to map out how life in contemporary systematics is lived and the concerns that animate its engagements with ICTs; but it cannot be other than a partial account.

The field site I explored I have come to call, in shorthand, "systematics as cyberscience." I am aware, though, of the ironies implicit in writing a book about these technologies as if they were all the same phenomenon while at the same time arguing for their disaggregation and attention to the specificity of occurrence. The approach that I have taken, lumping together ICTs as a singular object of study, inevitably involves definitional problems. At times it has not been clear which technologies to focus on, and whether a given technology raises any interesting and more general questions for my purposes. This concern is not solely confined to this particular technology. As Benschop (1998) describes, studying the trajectory of a scientific instrument can provide considerable historiographic problems as definitions and functions shift over time. Whether historian or ethnographer, it is a mistake to be too certain about the identity of the thing you are looking at, since by doing so you may miss out on significant strands of meaning-making that refashion the objects of interest. In taking together all ICTs I have effectively invented a category not often used by systematists themselves. I have done this as a means of exploring the ways in which the discipline has made this highly culturally significant category of technologies its own. It is not the idea, then, to say that ICTs always come together as a package, but to explore the specific and diverse ways in which they are deployed.

It is clear that conducting a credible study of systematics requires looking at the places where work in systematics is done and discussed. Within the UK at least, most of the systematics work is done outside of universities, and focused on institutions like the Natural History Museum in London and the Royal Botanic Gardens in Kew and Edinburgh, as well as numerous smaller museums. In chapter 4 I discuss some of the purchase on online systematics that was offered by going to some of these institutions and developing an understanding of the material culture that accompanies developments in online systematics. In order to find out how some of these activities made sense, however, I had to go beyond the apparently enclosed locations of systematics research offered by the herbarium, the laboratory, and the museum. As Blume explains: "The point I want to stress is that the material practice of science (establishing the scope and nature of a series of experiments, choosing or designing the instrumentation to be used, making the observations) and the kind of negotiations which experimentation involves (with professional peers, laboratory heads, funding bodies of various kinds) are inter-dependent" (1992: 93). Interviews with systematists in a variety of institutional roles and locations, together with explorations of documents produced by a wide range of practitioners of and commentators on systematics, provided an insight into this kind of interdependency and allowed exploration of diverse ways in which online activities are co-constructed with other aspects of the discipline.

The burgeoning array of systematics sites on the World Wide Web provides a significant space to observe what systematics is making of these technologies, to explore traces of the antecedent circumstances that have produced this sphere, and to capture the emerging qualities of that sphere itself. The World Wide Web has become a significant site of systematics. So too have electronic communications between individual systematists and between groups. Although I did not have access to the private email communications of individuals, any more than I had access to their private conversations and correspondence, I have been able to access and participate in open group forums online, and to use email as well as face-to-face interviews with systematists as research tools. I have made use of standard tools available to explore the Web: I used the Google search engine, aware that search engines cannot produce objective mappings, seeking nevertheless to use them to explore the readily available Web territories that have formed in contemporary systematics. In this I was assisted by the visualizations offered by the TouchGraph Google Browser (http://www.touchgraph.com/TGGoogleBrowser.html/), which relies on the Google facility to track down "related" sites. I make considerable use of this utility in chapter 6, tracking the landscape of institutions and initiatives that make up systematics as cyberscience.

A core site for exploration has been the online discussion list of the taxonomic community, Taxacom, and I both attend to the qualities of this list as a medium for disciplinary reflection and also use the list for its content, as it demonstrates features of the material and virtual culture of systematics and the beliefs of its practitioners. In exploring mailing lists I have used an ethnographic approach that relies on immersion and participation to develop understanding of their composition as cultural sites in their own right. I also use simple tools to parse the available listserv archives and draw on the kindness of the list owners in making list membership statistics available to me. The details of this approach are explained in chapter 5, which describes the emergent communication practices of systematics. In all of these approaches, I exploit the qualities of the Internet as a medium that bears traces of social engagement, but I remain conscious that not everything that people do online leaves a trace (Beaulieu 2005). My understanding of mailing lists and Web sites was underpinned by email interactions, face-to-face interviews, and site visits, which allowed me to develop a sense of the ways in which these pieces of online culture were meaningful within a disciplinary culture. The connective approach to ethnography relies on following strands of meaning-making across what may seem at first sight to be self-contained cultural domains. Rejecting the idea of a discrete virtual culture (Miller and Slater 2000) as an *a priori* methodological principle, I aimed to understand online contexts as they were contextualized and made meaningful by those involved.

This book is based on observations across a range of different contexts and through different media. Part of the work draws on existing histories of some of the botanic gardens, herbaria, and museums which are a prominent part of the material culture of systematics. I visited some institutions as a conventional paying visitor, as a participant in organized "behind the scenes" tours and presentations, and as a researcher interviewing members of staff. The visits to actual sites were, of practical necessity, limited to those that were geographically close, and hence the book emphasizes in particular the Natural History Museum, Chelsea Physic Garden and the Royal Botanic Gardens, Kew in London, the Millennium Seed Bank at Wakehurst Place (the country outpost of the Royal Botanic Gardens, Kew), the botanic gardens and plant sciences department of the University of Oxford, the Muséum Nationale d'Histoire Naturelle in Paris, and the Hortus Botanicus in Amsterdam. My interest has included the collections themselves, plus display materials, visitor information, and presentations by guides. I have also explored the virtual manifestations of these institutions, visiting Web sites and exploring online databases, and studying the academic literature relating to the collection and storage of specimens and the role of computers in taxonomy. The majority of these visits, online observations, and interactions with systematists took place between 2002 and 2005. In chapter 6 I also draw on ethnographic engagement with a particular project, ILDIS, from its early days in the 1990s. At that time I attended meetings, interviewed participants, and had access to both its public representations and its private files. Subsequently I have followed the progress of the project and its influence on new initiatives through the literature and via its Web sites. In this I have been considerably assisted by the very open cultures that such projects have often adopted, and the habits of presenting meeting schedules, minutes, and working documents on the Web.

In addition to the activities carried out explicitly in the service of this research project, there is also an autobiographical element. As I have been an enthusiastic visitor of natural history museums and botanic gardens for many years, this project has been prefigured by visits to botanic gardens and natural history museums in cities including Stockholm, Uppsala, Leuven, Oxford, and Cambridge, and to the Smithsonian Institution in Washington, D.C. As a master's student, my final project and dissertation was based on work at the Natural History Museum in London, constructing a database of British ferns. I was concerned with ferns as data and as sets of characteristics described in the literature, and although my direct immersion in the material culture of systematics was limited, I did develop a familiarity with the practices of the herbarium alongside which I worked. I had a similar experience as an undergraduate, when I undertook a summer job at the University Museum in Oxford and was detailed to wash fossil specimens for display. I learned that no matter how tatty an old label was, it was a precious part of the specimen that could help to interpret it in future and must be carefully preserved. I cannot claim to have been doing ethnographic research at the time. There is no doubt, though, that my sense of the material culture of systematics has been thoroughly shaped by these immersions in its institutions and its daily working practices. This book, then, is a methodologically eccentric historico-ethnographical autobiographically inflected multisited analysis of the material culture, political and institutional landscape, and communication practices of systematics seen through the lens of their intersections with ICTs.

It will have been clear from the introduction that this is not an ethnography in the style of some of the pioneering STS studies such as Latour and Woolgar (1986) or Traweek (1988), where the figure of the ethnographer as stranger acts as a means to question aspects of scientific life that might otherwise be taken for granted. I have a prior engagement and continuing closeness with the discipline I study that would make this "stranger" stance impossible to sustain. My background of insider experience of the commitments and working practices of systematics means that I cannot credibly portray myself as a stranger; and to do so would be to miss out on the considerable benefits that my background offers, in terms of my level of engagement with the details of systematics and the possibilities for enhanced interactions with practitioners. The level of prior engagement that I had helped to make the wide-ranging connective approach possible in a relatively short time, since I already had an awareness of the key concerns shaping the field and a sensitivity to distinctive aspects of its culture. There are, however, pitfalls for the insider ethnographer to guard against, and among them is failing to maintain a balance that preserves depth of engagement with the concerns of the field without losing the analytic edge offered by ethnographic styles of "othering" the subjects of the study (Dyck 2000). The drawbacks of an ethnography that is overly in sympathy with a particular perspective are well recognized, even though they balance against the difficulties an evident outsider may have in accessing and attending to the detail of social experience (Aguilar 1981). It is possible, then, that I have failed to defamiliarize aspects that a "stranger" ethnographer might have brought into question, or have had my descriptions colored romantically by my sympathies.

This is not, however, simply an insider ethnography, either, even though it has elements of that approach. I tried to get along with the systematists I encountered, but I did not aim to be one of them. Although I am also an academic, I am not a systematist, nor have I ever been dependent on that community for my professional advancement or recognition. It is to colleagues in sociology and STS that I turn as my peers, despite my considerable personal sympathies with the goals that systematics pursues and my concerns about the constraints it faces. My experiences as a sociologist contrast with my understandings of biology, and the comparison of these two very different academic fields gives me insights to interrogate the taken-for-granteds of each domain. As Aguilar (1981) points out, all cultures are diverse and fragmented and to suggest that anyone is either wholly inside or outside a culture is oversimplistic. Allegiances also change over time, trajectories vary, and ethnographers experience times of closer identification with the world of those they study mixed in with other times of a sense of distance (Coffey 1999). This may become particularly apparent in multisited ethnographies, which are, Marcus (1995) argues, characterized by a circumstantial activism enacted in shifting sets of relationships. In STS the topic of engagement with fields of study is a live issue, as concerns are raised about the apparent tension between radical constructivist sensibilities and productive engagement with domains of use.[1] We act in relation to the circumstances in which we find ourselves, but we also find ourselves altered in the process. Ultimately, then, the study I describe here is more a co-construction of systematics and myself than an objective portrayal of systematics.

While I have aimed to be fair, I have made no attempt to erase my complicity from this text, and it is inevitably a product of a diverse set of professional and personal affiliations. I have opted for a style of reporting that tells of the world as it makes sense to systematists rather than stressing repeatedly the contingent and constructed nature of their world. This decision stems, in part, from my closeness to this world, my wish to tell of the diverse pressures and constraints that permeate the efforts of systematists to do what they consider the right thing, and my concerns to produce a style of writing that science practitioners can engage with. In this my reporting style is somewhat different from the "ethnographer as stranger" style of the classic works in laboratory ethnography. My interpretations of how the world makes sense for systematists have a habit of blurring into accounts they offer. While my vocabulary of co-construction and imagining and my talk of branding lay a set of constructs across their actions that systematists would not use themselves, in many other respects my account is not so radically different from the accounts that systematists might give. An important part of my experience of systematics was that its practitioners were often highly reflexive and effective practical sociologists. To layer my own sociological structure too heavily across the top would be to do injustice to this practical sociological insight that systematists themselves possess. I have therefore aimed to tread relatively lightly with distinctions between insider and ethnographic accounts, while ultimately telling a story that only I could produce.

The following chapters of the book present different aspects of the study, and each chapter presents a more detailed description of the methods used to acquire data where this is called for. I have kept these detailed descriptions of method close to the point at which the data is used, in order to highlight in the relevant places the partiality and contingency of the account. Chapter 3 is somewhat of an introductory chapter, in that it presents a first attempt to map the diverse ways in which use of ICTs makes sense in contemporary systematics. The particular set of data on which the description is based gives the analysis an additional twist, since it consists of a report from the House of Lords Select Committee on Science and Technology focusing on the state of systematics. This chapter, then, not only introduces systematics, but paints the picture of systematics as a politicized discipline subject to considerable high-level attention. This chapter brings the autonomy of systematics into question, and although the report is a poor reflection of the complex dynamics of the policy domain, it gives a glimpse into the way that policy for systematics has seized upon the potential of ICTs to act as an agent of change. Subsequent chapters then separate out particular themes for exploration from the perspective of a richer set of observations from connective ethnography, looking in turn at the entwining of virtual and material culture, the emergent communicative practices, and the emergence of initiatives in ICTs as meaningful activities for individuals and institutions to engage in. The final chapter ties these separate chapters back together again, looking at the emergence of ICTs as a potent way of imagining the discipline and considering what kind of discipline is emerging from its co-construction with ICTs.

3 Computers and the Politics of Systematics

Policy makers are engaged in imagining the futures of the domains which come under their influence. ICTs have recently become a potent way to imagine the future of science, as demonstrated in the previous chapter. As sites of imagining, policy documents commenting on ICTs for science portray the present and future of scientific disciplines and offer mechanisms for moving forward. Mackenzie's (1996) principles for sociology of technology make clear that while not to be taken as straightforward prescriptions for action, these policy documents have a part to play in the dynamic field of technological innovation, as they become resources for understanding present and future actions and can become self-fulfilling prophecies.

In this chapter I begin the work of understanding these dynamics within one particular field of science. I focus on a report which contains some high-level pronouncements on the role of ICTs within systematics, and identify core themes that emerge from the co-construction of the technology and the science for this arena. This provides an understanding of some of the political concerns that suffuse everyday experiences of ICTs for systematics, which are then explored in more depth in subsequent chapters. The chapter begins by setting the scene within systematics, providing context for a discussion of the policy report that forms the focus of the analysis. Having explored four key themes that arise in this one report, a final section of the chapter looks at the history of these concerns within the discipline. This chapter therefore begins to establish that ICTs have not merely parachuted into the discipline as agents of change. These technologies have histories, they have advocates and opponents, and they have cultural meaning for those involved.

Introducing the Field of Systematics

The area of science chosen for this study is biological systematics, or taxonomy. This discipline provides the nomenclature that is used by the rest of biological science and, by producing biologically meaningful classifications, provides a predictive and comparative resource that can be used by other biologists to suggest hypotheses and directions for research and by conservationists to inform their interventions. Producing these classifications is a far from trivial task. Estimates of the number of species on earth vary wildly, but always produce a total that is vast in relation to the numbers of taxonomists available to describe and name them. Erwin (1997), as author of one of the higher estimates of total species numbers (30–50 million species of insects alone), describes the problems that this mismatch between the scale of the work and the practices for dealing with it poses for using knowledge of species diversity for conservation and environmental monitoring:

> The reasons [for the lack of use of beetle information in environmental monitoring] probably lie in the overwhelming numbers of species, individuals, and the ever-plodding course of traditional taxonomy. Potential users of data on beetles simply have to wait too long to get names; taxonomists have to wait too long to receive money to visit museums where name-bearing types are held; monographers take too long to produce documents with which users might identify their specimens by themselves; and specialists are reluctant to take on a large identification load for other scientists, such as ecologists and conservation biologists. (Erwin 1997: 29)

This excerpt illustrates some key preoccupations of systematics: the mismatch between the scale of the task to be done and the resources available to tackle it; and a problem of image which leads other biologists to distrust the ability of the discipline to address the problem. Insofar as the rest of biology requires names for the organisms that it studies and needs some guidance on their similarities and evolutionary relationships, biology as a whole is dependent on taxonomy. Erwin's depiction of taxonomy as "ever-plodding" marks a dissatisfaction with the working practices of the discipline that is believed to be widespread. This perception that all is not as it could be with taxonomy as a discipline in its own right, as a service to biology and as a vital component of efforts to understand and preserve biodiversity, permeates discussions of the use of ICTs in the field.

Modern systems of nomenclature date from the work of Linnaeus as published in the eighteenth century, and working taxonomists (and users of taxonomy) now have to grapple with complex synonymies built up by superseded and discredited classifications since that time. Although there are complex rules on the steps that must be gone through for a name to be considered valid, according to separate nomenclatural codes for animals, plants, bacteria, viruses, and cultivated plants, there is no final arbitration on whether a classification and the names that depend on it are accepted.

Acceptance is a matter of scientific judgment, and any consensus emerges de facto. Increasingly systematists work with DNA sequence data to add to more traditional techniques and use software to carry out analysis of phylogeny, resulting in hypotheses of evolutionary descent. They also face possibilities that developments in DNA barcoding might sideline traditional approaches to classification and nomenclature altogether. As we will see in more detail in future chapters, many new methods and new approaches to taxonomy have been incorporated into existing practices and institutions, while still retaining and building on a heritage of past taxonomic work and highly formalized practices.

There have been enduring concerns about the image of taxonomy (Vernon 1993). As this chapter will demonstrate, taxonomists continue to be concerned that other biologists do not respect or appreciate them and that users outside biology have little regard for them. The proposed solutions to this image problem have frequently involved the use of ICTs (Hine 1995). Subsequent chapters examine specific instances of ICT use in systematics in this context, in terms of the rationales that make them meaningful and their implications for working practices and outcomes. As a starting point, however, it seems useful to survey the territory. I would usually, at this point, begin mapping the key institutions and individuals able to comment on the current state of systematics, and would set about gathering a range of views through interviews and documentary analysis. Quite fortuitously in this particular case a recent report is already available, which gives a useful set of materials with which to consider the state of contemporary systematic biology in the UK as expressed by the most interested parties. The existence of the report in itself tells us a lot about the current status of systematics and the set of themes that emerge form a starting point for analysis of the recent history and current practices of ICTs in systematics.

In 2002 a report was published strongly recommending that systematic biology take steps to become a Web-based science. Species descriptions and interactive identification keys should be published on the Web, and increased funding should be made available to digitize collections of specimens and make taxonomic information available to people all over the world. The proposals are for a radical reshaping of practice to incorporate ICTs into every stage of taxonomic communication. Unusually, these calls did not come from within the community of systematic biologists. Instead, they were contained in a report of the House of Lords Select Committee on Science and Technology. This groups of peers (Lords Flowers, Haskel, Lewis of Newnham, McColl of Dulwich, Patel Quirk, Rea, Soulsby of Swaffham

Prior and Turnberg, the Earl of Selborne, and Baroness Walmsley) state in the summary of their report:

> We highlight the importance of digitising the systematic biology collections, which will both increase accessibility of these data and help to update the archaic image of systematic biology. We also suggest that the systematic biology community should consider exploring new ways of presenting taxonomic information, in particular through increasing the amount of information available in digital form via the world-wide web, and should consider updating the system of naming previously undocumented species. (Select Committee on Science and Technology 2002a: 5)

If, in the first place, it is unusual for the House of Lords to take such a detailed interest in the workings of a scientific discipline, it seems even more strange that their recommendation should so clearly point to a technological solution to the problems identified. This chapter examines how systematics came to the point of receiving such high-profile attention, and how ICTs were framed as the solution to its problems. An analysis of the report, including the evidence offered by a wide range of parties interested in the development of UK systematics, provides a set of themes for comparison with the history of predictions about and usage of ICTs in systematics. As suggested in the previous chapter, predicting the future and recommending ways of moving toward desirable futures are in themselves highly complicated activities which are connected with everyday practices in a variety of ways. They help give meaning to the everyday, and as Iacono and Kling's (2001) comments on computerization movements show, they can provide guidance on favorable directions for everyday practice to pursue while also often clashing with on-the-ground experiences of ICTs in action.

What on Earth? The Select Committee Report

The background of the Select Committee report lies in the spotlight recently placed on biological systematics by increasing political interest in biodiversity and conservation. High-profile events such as the Rio Earth Summit in 1992 and the Johannesburg World Summit on Sustainable Development in 2002 turned the conservation of wildlife into a matter of international politics. At the Rio Summit, the Convention on Biological Diversity (http://www.biodiv.org/) was signed. This agreement committed its signatories to undertake a variety of activities to safeguard the preservation of biodiversity, both in their own territories and by assisting developing countries. Among these activities was the need to survey and document the extent of biodiversity. The term "taxonomic impediment" was used to signal the lack of both available taxonomic knowledge and expertise needed to complete such a survey. It was argued that to be effective and to be monitored effectively, conservation requires that we know what species are present, and that in turn this relies on the work of taxonomists. The Rio Summit, and the Convention on Biological Diversity, thus meant that some 157 countries publicly committed themselves to supporting the work of taxonomists. It is hard to imagine another branch of science, outside medicine, gaining such widespread endorsement.

The Rio summit did give taxonomy a high profile, but this is not to suggest that all has been easy for taxonomy since Rio. The 2002 report was an update on a 1991 Select Committee report (Select Committee on Science and Technology 1991) on the state of systematics in the UK. The 1991 report found, in brief, that systematics was strategically important but underfunded for the task it faced. The 2002 update found that the 1991 recommendations for funding of taxonomy had not broadly been met: indeed, the increased political focus had, some felt, imposed new demands on systematists without increasing funding to match. By becoming politically high profile, taxonomy has also been forced to make public its concerns and to justify what it does with the funding it receives. Political prominence has, then, been a mixed blessing; but on balance at least the existence of these reports is a marker of the recognition of the significance of the discipline. In the report, and the contributions made to its deliberations by written and oral evidence, we have a remarkable set of data in which a scientific discipline is cross-examined about its goals, its practices, and its organizational and funding regimes.

The report is a valuable set of data for a sociologist of science interested in exploring the key players in a field: here they step forward to position themselves and give an account of what they see as the problems and potentials of the discipline. Clearly these statements are delivered in a highly politicized arena, so as a representation of the way the field "really" is they must be treated with some caution. Interviewees in various domains helped me to realize that it is not straightforward to draw a clear boundary between policy makers or commentators like the Select Committee and science practice as represented by the institutions. The major institutions saw the occasion of the report as a key opportunity to get their perspective across, hoping to set the agenda through educating the Select Committee on the appropriate ways to see the situation. For each of its investigations the Select Committee appoints a temporary specialist adviser, and in this case the specialist was a representative of the Royal Botanic Gardens, Kew. Interviewees had positions on how far the report did a good job in putting forward the concerns of the systematics community. Rather than an objective portrayal of the field, therefore, the report is the upshot of a complex set of dynamics and subject to diverse interpretations by different audiences. It is also a carefully crafted artifact taking the conventionalized form of a Select Committee report, and it is directed at a particular audience, most directly the government of the day to whom the Select Committee in the House of Lords is positioned as an adviser, but whose advice the House of Commons is by no means obliged to take. The Select Committee report is thus more a ritualized form of political discourse rather than the policy process per se.

With these provisos in mind, however, and recognizing that any interview, statement, or document that I might have elicited from informants would have also been delivered in a context of political self-awareness, this is as clear a map of contemporary UK systematics as could be hoped for. The form of the Select Committee report dictates that it should preserve all of the evidence that was offered during the investigation together with the recommendations built on that evidence. It therefore contains multiple voices, even though it obscures the array of additional voices and concerns that surrounded its production but did not take the shape of formal evidence. In the rest of this section I examine the main report and the volume of evidence given to the Select Committee for themes that emerge in the representation of ICTs and their role in contemporary systematics, looking to explore both the direction in which systematics is being steered and the response of practitioners of systematics to the steer they are experiencing.

When the report was first commissioned, an open invitation was issued to give evidence to the committee. The call invited contributions on the following questions:

> How has the organisation of and funding for systematic biology in the United Kingdom changed since 1992?
>
> What, if any, are the changes required in this area to enable the United Kingdom to meet its policy aims on biodiversity? (Select Committee on Science and Technology 2002a: 29)

Evidence was offered by a set of institutions and organizations representing key positions in British taxonomy. Learned societies, museums and botanic gardens, government departments, funding councils, and academic departments were represented, together with prominent individuals in the field, and the occasional less prominent individual responding to the open call. A selected few were invited to give evidence in person to the committee. In addition, a seminar was organized at the Natural History Museum in London to introduce the current state of biological systematics, and the committee also visited Kew Gardens for an introduction to the work of its herbarium, library, and Jodrell Laboratory. The written evidence and transcripts of oral evidence are available to accompany the final report. Also available in the public domain are the transcripts of a House of Lords debate on the report, the government's response to the report, and a set of memoranda from interested parties commenting on the government response, plus the Select Committee's commentary on that response.

Despite the fact that the report's remit did not focus explicitly on ICTs, commentaries on these technologies occurred throughout the report. As I read it I marked each place where ICTs were mentioned with a sticky paper tag, prior to transcribing these portions for thematic analysis: figure 3.1 shows the result, demonstrating in visible form just how dense the report's reliance on ICTs was. I subsequently looked at the places where references to ICTs occur in this evidence, and at the qualities attributed to these technologies in terms of their abilities to solve (or occasion) problems. As I have done previously in rather different circumstances (Hine 2000), I aimed to use this set of materials to examine the construction of ICTs as cultural artifacts. The analysis is presented as a set of broad themes that arise

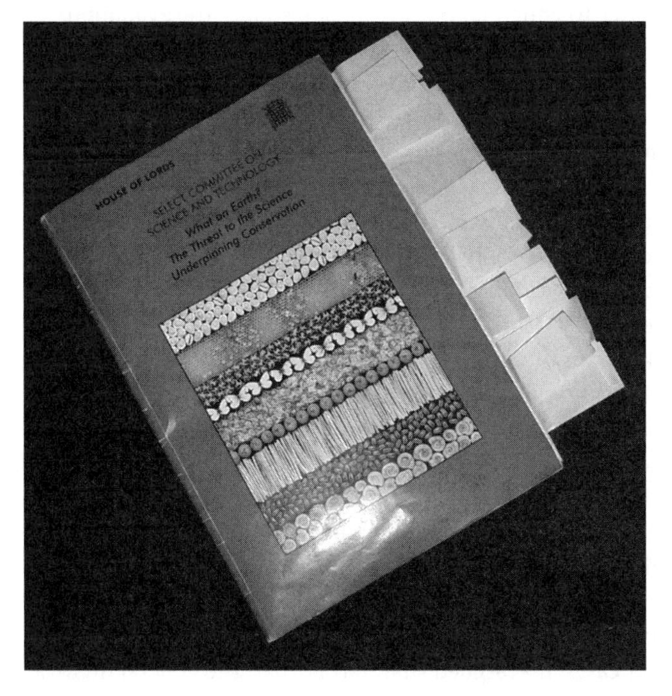

Figure 3.1
My copy of the *What on Earth?* report (Select Committee on Science and Technology 2002a) with paper tags marking each mention of ICTs.

repeatedly in evidence and locate particularly problematic, contested, or influential issues in the application of ICTs within systematics. The themes were arrived at through a process of identifying segments of the text that mentioned some aspect of ICTs, and then iteratively assigning and aggregating categories until four relatively stable themes which arose repeatedly throughout the report and evidence were revealed. The themes are not directly those of the participants themselves: they are a sociologized version of issues that do concern participants, translated into terms that have more currency in sociology of science and technology.

Material and Virtual: The Adequacy of Digital Specimens and the Audiences for Availability

Probably the key concept running through the report, as far as ICTs are concerned, is the need for digitization. Repeatedly in the main report and in individual pieces of written and oral evidence, it is stressed that systematics holds a wealth of valuable information in its collections of specimens and its literature, which simply must be digitized and made available. The coming of the Internet, and particularly the World Wide Web, is seen as mounting a challenge to which taxonomy must respond: the availability of the technical means provides the imperative to make use of it. The main problem identified is that museums and herbaria hold vast collections of specimens that need to be digitized in order to be made available, and that this will require funding and effort.

In the evidence given to the Select Committee, it is stressed over and over again that digitizing of specimens is a question of availability. In its material form, the specimen is held to be relatively unavailable. The equation of digitization with availability did, however, gain a closer examination in some of the oral evidence. First, the ability of users to distinguish reliable information from rubbish was questioned, and second, the ability of users to ask the right questions was doubted. When asked if digitized information was available to users in universities, the representative of the Joint Nature Conservancy Council replied: "Yes. There is a much wider access to these data in universities and other educational institutes. The problem is that there are insufficient course modules to enable people to start asking the right questions and to begin the process of understanding how systematic biology operates" (Select Committee on Science and Technology 2002b: 110). A distinction was here made between simple availability and meaningful use. In other cases, it is argued that, rather than the general digitization of data, taxonomists need to develop particular products, aimed at specific audiences, and place these on the Internet. This distinction was, however, rare. In the majority of cases digitization, or placing databases on the Web, was represented as in itself making taxonomic research and data globally available (more in the next section on the political connotations of global availability).

The digital specimen then appears to be the ideal specimen—instantly available to a wide range of audiences—and the only sticking point is the need for funding to carry out the "one-off" task of digitization. Another advantage of digitization that was proposed, as far as the collection holders are concerned, may be that pressure is taken off the systems for loan of specimens. Where a digital representation can be portrayed as adequate there would be no need to send off the material specimen to a distant researcher. The collection holders are, however, at pains to point out that the digital representation does not render the material collection redundant (possibly foreseeing the need to head off suggestions that the costs of ongoing curating of material specimens could be saved). Written evidence from the Department of Plant Sciences, University of Oxford, made this point: "However, computers do not make the collections themselves redundant. Whilst one can digitize an image of a herbarium specimen and make it and the associated specimen label data easily accessible to a wide range of users on the World Wide Web, research that involves approaches such as anatomy or molecular analysis must still rely on the physical specimen" (Select Committee on Science and Technology 2002b: 57). The importance of physical specimens in this perspective lies in the impossibility of predicting in advance, and thus capturing in digital form, all the qualities of a specimen that might interest future taxonomists. A particular category of specimen, the type specimen, holds a special significance in taxonomy. When one describes and names a new species, one nominates a particular physical specimen as the type, to which the name is formally attached. In any future revision of the definition and boundaries of this species, the original species name stays with the type specimen. This is seen as almost guaranteeing that a future taxonomist reexamining the definition of a species will need to see the type specimen in physical rather than virtual form. Physical specimens are seen as carriers of potential future information, as yet unrecognizable, in a way that digital specimens are not.

The evidence therefore enacts a complex dance in representing digital specimens as both universally available and at the same time not adequate in themselves for all systematic purposes. The self-evidence of the need for digitization is taken as a point of leverage for more funds, since it augments rather than replacing the traditional activities of curating specimens. Holders of collections stress that digitization, while essential, lies outside the activities for which they have traditionally been responsible and been funded. The report as a whole, in equating digitization with availability and in taking as self-evident that availability is a good thing, makes that case for increased funding appear incontrovertible (although this turns out not to be the case in the government's response to the report).

The equation of digitization with availability also takes on additional significance when we consider the audiences for availability. Systematics has been deeply concerned not just with its direct audiences, that is, those who make use of its information, but also with the broader audience for its work as a discipline. Systematics, and in particular taxonomy, feels itself to have an image problem. This problem was the focus of considerable attention by the Select Committee. It was felt that although the poor image represented a misunderstanding of the importance of systematics, and led the discipline to receive less respect and less funding than it deserved, this image was also to some extent the fault of the discipline, and therefore some responsibility for rectifying the situation lay with its practitioners. One of the committee's nine recommendations was as follows:

> We recommend that the systematic biology community, especially via the Systematics Association and the Linnean Society, should continue to increase efforts to demonstrate the relevance and importance of systematic biology. This should have the effect both of improving its profile to funding bodies and of making it more attractive to potential professional taxonomists and volunteers. We also hope that systematic biologists who are members of learned societies, such as the Institute of Biology and the Royal Society, will use their influence to promote the discipline. (Select Committee on Science and Technology 2002a: 6)

That the image of systematics was indeed a problem ran through the committee's deliberations: some time was given in the oral evidence to discussion of whether "archaic" or "arcane" were the more suitable term to describe taxonomy's current practices! In this context, the use of ICTs figured as an image of modernity that could help to correct taxonomy's poor image and be seen to bring it up to date.

The audience for digitization explicitly included not just direct users of information, but also a broader imagined audience alert to the technologies that the discipline was using and astute to the meanings carried by those technologies. Repeatedly in the written and oral evidence systematics was urged to make use of modern technologies in order to improve its image. Plantlife, a wild plant conservation charity, suggested that use of information technology would appeal to the young in particular: "We need a web-based resource where any group's taxonomy is placed on the web and then future revisions or additions would also take place on the web.

The implications of this move would obviously be great, but would surely change the image of the discipline and make it a more attractive career prospect for young scientists" (Select Committee on Science and Technology 2002b: 55). This perception of technology use as affecting disciplinary image was clearly influential for the Select Committee. In the summary of their report, two functions of the digitization of data were given equal prominence: "We highlight the importance of digitizing the systematic biology collections, which will both increase accessibility of these data and help to update the archaic image of systematic biology" (Select Committee on Science and Technology 2002a: 5).

Already, then, we have some insight into the complex arena into which ICTs are inserted and the symbolic qualities that are significant for the systematics community. What these technologies can do is worked out on a highly charged and contested field, on which funding regimes, institutional and disciplinary images, and sets of rights and responsibilities can depend. Systematics has also, however, become a much more explicitly political domain, and this too affects discussions of ICT use, as the next section demonstrates.

The Complex Political Geographies of Systematics

The history of taxonomy is to a large extent a colonial history. In the past, nations with colonies in other parts of the world would exploit the natural biodiversity resources of those countries as unquestioningly as they might draw on their peoples and their mineral wealth. The history of the major taxonomic institutions is tied up with traditions of travel and collecting in far-flung parts of the world. Viewed from the present day, this situation is highly problematic. The politics of biodiversity involves clear inequalities. Both specimens and the expertise to study and interpret them tend to be located in the richer nations, while the economically poorer nations are often rich in biodiversity but lacking in both expertise and the raw material, in terms of a heritage of physical specimens, to take charge of their own natural resource. It is this context that animates current thinking on biodiversity information.

At the Rio Earth Summit in 1992 the question of biodiversity and the need to conserve species and habitats entered onto the world political stage in an unprecedented fashion. At the summit and subsequently, the Convention on Biological Diversity was signed by 157 countries. The politics of biodiversity were explicitly written into the convention, in that it attempted to redress the imbalances between richer nations who were so often the holders of expertise, and the poorer nations who were biodiversity rich. Taxonomy was recognized as underpinning attempts to conserve biodiversity. Indeed, the convention enshrined the notion of a "taxonomic impediment" to biodiversity conservation, caused by lack of basic taxonomic data and lack of access to what data there was. Signatories to the convention were explicitly required to share what taxonomic information and expertise they had. The notion of "data repatriation" was used to stand in for meeting this requirement and is cited often in the Select Committee report, both in the summary report and in written evidence by interested parties. By digitizing collection information and placing the results on the Internet, the requirement to share taxonomic information is considered met. The possibility of sending physical taxonomic specimens, other than on temporary loan, does not arise.

The requirement for sharing of taxonomic expertise has, according to the report, largely been met in the UK by projects carried out under the Darwin Initiative. This scheme, developed in 1992 and administered now for the UK government by the Department for Environment, Food and Rural Affairs, has allocated approximately £3 million per year through a range of activities aimed at fulfilling obligations under the Convention on Biological Diversity. Projects funded under this initiative were praised in this report for their activities in sharing information and expertise. An exemplary project cited in the report is described thus:

> The Natural History Museum, the Plymouth Marine Laboratory and the Kasetsart University in Bangkok and the Ministry of Fisheries, Thailand collaborated to explore the potential of the world-wide-web as a tool for exchange of taxonomic information between biodiversity researchers. The project has enabled researchers to share information about polychaete worms, in order to determine whether specimens found in various places are the same or different species. Polychaetes are segmented worms and are very common marine organisms. Taxonomic information on them is used to monitor environmental quality. This Darwin Inititiative project has contributed to developing a high quality base of taxonomic information for use by marine conservationists. (Select Committee on Science and Technology 2002a: 15)

The project is presented as exemplary in its focus on an environmentally important group of organisms, its internationally collaborative organization, its aim of sharing information, and its use of the World Wide Web. The Darwin Initiative, then, through the notion of "data repatriation," is portrayed as drawing on ICTs as a way of meeting the UK's obligations under the Convention on Biological Diversity. The value of this source of funding is stressed in the report as also of general benefit in the goal of digitizing collections, in addition to any role in redressing the colonial legacy of systematics. Projects to develop Internet-based resources, particularly where they also involve training programs with economically poor but biodiversity-rich nations, seem appealing for their long-term availability in a funding climate where money for any kind of systematics research is seen to be extremely scarce. This is, however, a matter for interpretation by applicants to the fund, since there is no explicit steer in the rubric of the Darwin Initiative projects toward use of ICTs.

A further dimension of the complex political geography of systematics is provided by the Global Biodiversity Information Facility (GBIF). This initiative of the OECD Megascience Forum Working Group, supported by the parties to the Convention on Biological Diversity, has the remit of producing interfaces to biodiversity databases, including taxonomic databases, which will allow for relatively seamless searches across different sources of data (in GBIF terms, "an interoperable network of biodiversity databases and information technology tools"). This initiative and the UK systematics community's involvement in it were frequently used in submissions to the Select Committee to stand in as evidence for the willingness to embrace modern methods and make information available. Participation was therefore both globally important, in terms of meeting requirements as a signatory to the Convention on Biological Diversity, and locally important in terms of demonstrating, within the UK funding context, a willingness to be modern, to undergo change, and to embrace up-to-date technologies. Conscious of the UK funding context, many of the authors of written evidence to the committee were at pains to point out that while they were willing collaborators with GBIF, it tended to deal only with existing databases and neither funded nor addressed the question of basic data input of information from taxonomic collections. In stressing participation in GBIF as a self-evidently good thing and a national obligation, holders of collections reiterated the case for increased funding to digitize their own collections.

Automation and Expertise, Ease and Difficulty

The previous section laid out the extent to which initiatives to create taxonomic databases and to place systematic information on the Internet within the UK take place within a complex political geography. Local taxonomic practices, including the use of information technology, are played out in a global context. These local practices are also historically situated, however, in that the technologies that the discipline uses and/or is willing to use are shaped by ideas of what it has done in the past, what is easy and what is difficult, what is routine and what is innovative. As we have seen, there is considerable purchase for systematics in being seen to be innovative in its use of technologies. Some pieces of evidence to the committee pointed out instances of technology use that were already in place and could be considered to show how far the discipline had already moved. That the use of computers for calculating potential evolutionary relationships was established practice was pointed out by a representative from the Natural History Museum: "Many taxonomists today are primarily interested in relationships (phylogeny) of groups, and in estimating phylogeny using computer programmes based on character data. Major sources of characters in today's systematics are DNA sequences" (Select Committee on Science and Technology 2002b: 34). Acceptance of computerized methods could also be used in support of the idea that, left to conventional scientific practices, innovation could happen in systematics:

> Systematists and evolutionary biologists increasingly agree on accepted methods, available as computer applications for estimating systematic relationships from gene sequence or morphological data; David Swofford's PAUP is now the almost universally accepted route for estimating the tree of life from genbank. It is important to grasp the significance of now having accepted international protocols based on sound science. Evidence from Professor Paul Harvey FRS. (Select Committee on Science and Technology 2002b: 26)

Paradoxically, in demonstrating that they already use computers in their work in a routine fashion, systematists at the same time demonstrate that they are up-to-date and yet push the boundaries of what they can be expected to do. If use of computers is already routine in this area, then one might ask why it cannot progress still further. These strategies of highlighting existing computer use are therefore somewhat risky in a context where one is arguing for extra funds.

The distinction between routine work and innovative work was played out in representations of funding regimes. Although no clear sources of funding for the work of entering taxonomic information into databases were identified, it was suggested that money for developing innovative technologies was available, referred to by a representative from the Biotechnology and Biological Sciences Research Council as funding for "software and data bases; generic technical development" (Select Committee on Science and Technology 2002b: 164). The Royal Society, in its submission, suggested that the focus had been too much on innovation in technology: "The funding for electronic databases has been mainly technology driven and the information to populate these databases is sparse" (Select Committee on Science and Technology 2002b: 140). Identifying what is innovative and what is routine is therefore perceived to be of significance in negotiating a way through funding regimes. It is also, of course, important in representing the discipline as highly skilled and valued. To be seen as consisting wholly of routine work is problematic for the status of a scientific discipline, and certainly problematic where routine work is also likely to be seen as ripe for automation.

Some future possibilities for automation were presented in evidence to the Select Committee. The Royal Botanic Gardens Edinburgh suggested that automating routine identification of diatoms would help to overcome the "taxonomic impediment" to conservation of biodiversity. The Linnean Society approved of the idea of "automatic and on-line identification systems that enable non-specialists to make identifications" (Select Committee on Science and Technology 2002b: 126). More extreme versions of automation were produced by nonsystematists. Notably, the representative of the Office of Science and Technology suggested that this kind of project would have a strong fundability:

> So the whole question of taxonomic research today is in development where exciting things can happen, where visionaries could take it forward using modern biology methods, molecular biology as a basis, using modern instrumentation and, above all, using modern IT techniques, so instead of sitting examining with a microscope every specimen that comes your way and determining in fine detail what the structure is before its taxonomy is discovered, would it not be smart to develop techniques that could automate that process? I believe that the research councils would respond properly within their tensioning [*sic*] process to such proposals. So individual scientists are required as visionaries to come forward and take the field ahead. (Select Committee on Science and Technology 2002b: 149)

Where the possibility of automating various tasks associated with taxonomy, such as "routine" identification, was suggested, systematists offering evidence to the committee were often careful to point out that this did not remove the need for trained taxonomic experts. A University of Cambridge representative made a distinction between expert taxonomic work and the "service role" that could be automated:

> In not very many years from now, the preferred way of identifying specimens of "difficult" groups will be by obtaining a short DNA sequence. This is already beginning to happen for some groups of applied significance (eg fly larvae in forensic cases). To make this more widely possible, our long term aim must be to establish databases that correlate a DNA sequence tag with each classically described species. Once this has been done, specimens can be attributed to "existing species," or recognized as new to science, without huge taxonomic expertise. Of course, precise definition of species boundaries, the formal description of new species, taxonomic revision, and the generation of keys, will still require expert taxonomists who look at the animals- but the service role of such taxonomists will surely be replaced by

automated sample analysis over the next 20 years. (Select Committee on Science and Technology 2002b: 3)

In arguing for the value of their discipline and its continued funding, systematists are therefore representing their work and its relationship with technology on a public stage, and are involved in local and global politics. The demarcations they draw between routine and expert work, between physical and virtual specimens, between the various audiences for taxonomy, and between new and old technologies are highly consequential. Whether and in what ways taxonomy moves toward and is seen to move toward cyberscience will be played out on a highly politically charged and complex field.

Evocative Objects and the Branding of Initiatives

Recent research in science and technology studies points to policy making in the sciences as an arena of skilled rhetorical construction. Within this perspective, statements about the role of scientific disciplines and the need for research are shaped around attempts to make compelling visions of the future. Even the naming of disciplines can be seen as a rhetorical choice, as Hedgecoe argues: "This article's central claim is that rather than simply representing an area of research, the term *pharmacogenomics* can be seen as a rhetorical device used to gain support among policy makers and funders for particular research topics and technologies. By tapping into the interest and "hype" surrounding the word *genomics*, pharmacogenomics links into a number of future scenarios about the impact of genomic technology on health care systems and society as a whole" (2003: 513). For the purposes of the current discussion, it is important to note that the picture of systematics, or taxonomy, or the research underpinning biodiversity[1] that is given by the Select Committee report and the contributing evidence is far from neutral. The work that is being done, as described in the previous three themes, is aimed at presenting a compelling future vision that will prove attractive to funding bodies and potential recruits to the discipline, and improve the image of this kind of work among other biologists. In the case discussed by Hedgecoe, the shift from pharmacogenetics to pharmacogenomics was rhetorically important in allying the field to current interest in a set of future visions revolving around genomics. In the case being discussed here a different alliance is forged, between systematics and the Internet. The problem of systematics, as presented within the Select Committee report, is partly one of image. Systematics simply evokes the wrong kinds of qualities to be attractive to potential funders and possible future systematists. To address this problem, another evocative object with altogether different connotations is proposed: the Internet.

The term "evocative object" was coined by Turkle (1984, 1995, 2007) to capture our use of objects for the images they conjure up: some objects, such as the computer, carry connotations that frame the way we think about ourselves and the world around us. In Turkle's formulation, it is not entirely clear whether the objects carry some inherent evocative quality. I adopt the term "evocative object" here to mean an artifact that has acquired, for a particular set of users, a particular set of imagery. The evocation is not in the object so much as in the co-construction of the object with a particular domain, but this construction alludes to meanings that the object carries in other domains. The way that an object is imagined in one setting often draws on the way that other communities imagine similar objects. In this particular case, the authors of the House of Lords Select Committee report appear to see the Internet and the Web as objects that evoke desirable qualities of being modern, accessible, and user friendly. The evidence presented to the committee, and the report itself, therefore seek to develop desirable qualities in systematics by association with the desirable qualities of the Internet and the Web. The success of this strategy is not guaranteed, since it cannot be assumed that readers will associate the same qualities with the Internet that the Lords committee does. For some it might be time wasting, unfriendly, dangerous, frivolous, and so on—not qualities with which one would wish to be associated. The Internet does, though, appear to be a desirable technology at least for the group submitting evidence to the committee, if we consider that uses of the Internet and the Web are advanced by many groups submitting evidence as proof of their willingness to embrace new technologies and engage with users.

In the vast majority of mentions of the Internet, the equation of presence on the Web with positive effects is accepted without question. As the United Nations Environment Programme World Conservation Monitoring Centre portray it, the existence of these technologies mounts a challenge to which systematics can respond: "There are opportunities to make greater use of innovative and creative technologies in describing species and their communities, building on the UK's traditional strengths in systematic biology. The Internet provides important and largely under-utilised opportunities to share data on systematic biology worldwide, to the benefit of many nations" (Select Committee on Science and Technology 2002b: 113). That the Internet is embedded as a routine technology of information access is affirmed in many sources of evidence, including the personal testimony of the President of the Linnean Society: "I think there is such a natural tendency these days, even for people of my age, to go to the web to find information—we get very used to it. In serious science this is where it probably has to go, particularly when you bear in mind the ease of putting good illustrations on the web" (Select Committee on Science and Technology 2002b: 132).

Implicit in most of the evidence, and explicit in some places, the evocations of the Internet are foremost availability and accessibility of information to the widest conceivable audience, together with being up-to-date, forward looking, innovative, and creative. The main objection made is the obstacle which the costs and effort of digitizing existing collections place in the way of achieving the benefits of Internet accessibility: this point is made repeatedly by the curators of collections. As the notes on the seminar held for members of the Select Committee at the Natural History Museum have it: "Communicating the information from systematics to its many audiences: the Internet now offered the ideal vehicle for this and the necessary infrastructure was in place but putting the vast array of information contained within collections on the web presented a major one-off task for which special funds were needed" (Select Committee on Science and Technology 2002a: 34). Beyond this point about the costs of digitizing, only a few mentions are made of negative connotations of Web-based systematics. In one place the report mentions a problem of sustainability, attributing this to concerns raised by members of the systematics community: "There are a number of concerns about a primarily web-based science. For example, it has been argued that at present web-sites do not survive for long: few are properly maintained and many soon become out-of-date and redundant" (Select Committee on Science and Technology 2002a: 23).

A further concern raised about the Internet is the extent to which the information held there is reliable. This concern forms the topic of an exchange between the representative from the Joint Nature Conservation Committee and the Chair of the Select Committee:

McLean, JNCC . . . I think sometimes people get a little drawn into the world of information on the Internet and imagine somehow it is all very real. I am afraid there is an awful lot of rubbish out there as well as an awful lot of very sound information.

Chairman So you have some real concerns about the quality of that information?

McLean, JNCC Not with regard to GBIF and the Global Taxonomic Initiative, no. It is more widely that I am concerned. What I am saying is that the real challenge, as these processes expand and extend and get more and more contributors coming in, is actually going to be verifying the sources and understanding whether or not you can rely on them. So quality control, quality assurance is essential when building these big global enterprises. (Select Committee on Science and Technology 2002b: 109)

Here the concern that Internet landscapes can be a seductive terrain of unreliable information is raised. The Internet is portrayed as a challenge to which people may simply not be equipped to respond. At the same time that this concern is raised, however, a separation is made between this general risk and the specifics of particular initiatives in sharing of taxonomic information. Later in this chapter I will have more to say about the branding of initiatives. For now it is sufficient to note that the specter of alternative connotations of the Internet is raised but rapidly sidelined as not a genuine risk in the case of well-specified initiatives which can act as a guarantor of information reliability (or as what we might think of as a trusted brand).

Another evocative object that figures in the evidence to the Select Committee is genomics. Just as Hedgecoe (2003) found, association with genomics was considered to be and presented as a good thing. The possibility of a future DNA-based and largely automated identification process was raised in some evidence, in most detail by the University Museum of Zoology, Cambridge. In other instances, rather than a compelling future vision, the quality of genomics that was invoked was its vision and its ability to organize disparate groups of individuals into a large-scale collective initiative, and to mobilize the resources needed to address it. This was presented by Lord May of Oxford as a mode of organization to which systematics should aspire: "But such a hi-tech revolution will involve a completely different cast of characters, the kind of people who gave us the gene sequencing machines. We need people to be thinking ahead to that and getting together and putting proposals together in the same imaginative way as people like Walter Bodmer and others did to start on the human genome 20 years ago" (Select Committee on Science and Technology 2002b: 145). A "high-tech revolution" is thus seen as socially innovative as well as technologically innovative, and as requiring a vision to drive it. Genomics provides one model for shaping such innovation, or at least the encouragement of demonstrating that it can be done. Genomics is, however, only a minor player in the field of evocative objects drawn by this report. ICTs remain the key objects through which the possible futures of systematics are articulated.

On a rhetorical level, then, the Web and the Internet are seen in the report and by the contributors of evidence largely as evocative objects which bring a certain glamour and an image of modernity and accessibility to systematics. In Bowker's (2000: 656) terms, they are seen as "charismatic technologies." Where the main report is concerned, there is no need for an analyst like myself to stage a revelation of hidden interests, since the authors are extremely clear that the connection between systematics and the Internet is a strategic alliance for image purposes. For others submitting evidence, however, the link between Internet and image is less explicit. Partly this may be because the tenses are different: the Lords' report frames its advice on what should be done in future, while the systematics institutions giving evidence speak about the Internet in terms of what they are already doing. The two contexts provide for different ways of presenting the Internet, with the context of ongoing use being more about functionality than image. In both contexts, though, there is some attention to the evocative nature of the Internet and particular qualities of the Internet that are to be lent to systematics. This sense of evocation works at the broadest level of a technological and communications strategy. Much of the report, however, and particularly the evidence of the systematics institutions, focuses on the level of particular initiatives. Here we also see evocative objects in operation, but with a difference: through the branding of initiatives, systematics practitioners are attempting to portray themselves and what they do in a particular positive light.

By branding, I mean to express the extent to which activities within the field of systematics, and particularly the application of information technologies in systematics, are given public identities through names, acronyms, and publicity activities. Branding is a complex and indeterminate process, as Lury (2004) describes, yet its upshots can be potent objects. Ogden, Walt, and Lush (2003) describe the importance of branding in the policy process, showing how the profile of a particular policy initiative in the treatment of tuberculosis was marketed through a process of branding with a catchy acronym. In a similar fashion, projects or sets of activities within science increase their visibility to funders and to potential participants or users by developing a brand identity. Two examples occur with some frequency in the report: the Global Biodiversity Information Facility (GBIF) and the National Biodiversity Network (NBN). Although there is no direct piece of evidence from the GBIF, this project is mentioned in evidence by leading systematics institutions, by the Department for Environment, Food and Rural Affairs, United Nations Environment Programme World Conservation Monitoring Centre, and by learned societies and funding councils, and it is clearly oriented to as a significant entity by these key players in the domain of systematics. GBIF plays a part in the complex political geographies of systematics, as described above. It is a key component in the infrastructure envisaged to overcome the taxonomic impediment (Bowker 2000; Casey 2003). It also represents a major brand in contemporary systematics. As a brand, it symbolizes qualities beyond the merely functional: to buy into the brand is to buy into a particular vision of systematics, its organization and its duties. The importance of GBIF as a brand relies on the facts that it is conceived as global and that it is aimed not at producing a single database, but at providing for interoperability of databases. These qualities are seized on in the Select Committee report, which makes active participation and leadership in GBIF a key component of its recommendations for UK systematics.

A further brand that figures heavily in evidence and in the main report is the National Biodiversity Network (NBN). Like GBIF, this is an initiative to provide a common gateway to dispersed databases, this time within the UK. NBN again features in evidence as a notable brand, to which systematics institutions and funding bodies ally themselves. The NBN brand stands for accessibility of information and for cooperation between different institutions. It also stands as a specifically national response to global concerns. NBN and GBIF are presented largely as natural responses to ongoing pressures. The main report of the Select Committee describes NBN as "one example of increased awareness of the importance of digital information" (Select Committee on Science and Technology 2002a: 20). In this formulation, "increased awareness" is akin to an externalizing device (Potter 1996) that renders the situation as one of passive response to an independently occurring process. Although the importance of digital information is arguably constructed through NBN and similar initiatives (and as we have seen, there is little detailed idea of how digital information will be used and by whom), this formulation casts the brand as occurring subsequent to the need, rather than driving its identification. The brand is co-constructed with the environment it inhabits so as to position it as a natural response.

In evidence, numerous databases, software products, and Internet projects are mentioned by individuals and systematics institutions. It is made clear that an awful lot of work in systematics and biodiversity information is already underway, but none of these initiatives seems to have the qualities to catch the Select Committee's attention in their depictions of a desirable future for systematics. In their recommendations, only GBIF gains specific attention. GBIF appears particularly attractive as a transinstitutional and international initiative, which is relatively new (and therefore as yet unsullied by experience or disappointed hopes, as is the case for some previous brands) and promises to unite dispersed databases. Above all, this is a grand and large-scale initiative, which appears to have appeal for the Select Committee beyond small-scale efforts within a particular group of organisms or an institution. The goal of a wholesale revision of practice is an attractive option within the world created by the report.

The factor with the most influence on the recommendation of the Select Committee that taxonomy should strive to become a Web-based science was the work of one Professor Charles Godfray. Godfray, a population ecologist rather than a taxonomist, had published a paper in the newsletter of the Royal Entomological Society, recommending wholesale change in taxonomic working practices (Godfray 2002a). His suggestions, framed as ways that systematics might become more attractive to funders (including sponsors from the commercial sector), involved the intensive use of ICTs to revamp the process of assigning and communicating taxonomic nomenclature. The paper explicitly recommended the use of these technologies to improve the usability of taxonomic information and its image to nontaxonomists. Godfray aimed his recommendations not simply at databases to communicate taxonomic results, but at transforming the basic practices of taxonomic work itself, at least insofar as rules for assigning nomenclature were concerned. This paper was attracting attention in the systematics community and beyond at the very time most of those concerned were preparing their responses to the Select Committee's call for evidence. Thanks to its timing, this vision of taxonomic future had a far greater impact than even its author would probably have expected. There have been many calls for the revamping of taxonomy through the use of ICTs over the years (Hine 1995), each time focusing on the perceived problems of the discipline at the time and finding appropriate solutions in technology. Few, if any, have achieved this kind of public prominence. Godfray's paper provided a packaged "vision" of a future Web-based systematics at just the time when the Select Committee were seeking to find a way through the complex territory mapped by the evidence submitted to them.

It is significant that Godfray's (2002a) paper focused on what must be done if more taxonomy is to be funded. The explicit aim of the paper is to respond to calls from taxonomists for more funding with a clear outline of what forms of proposal would be fundable. The fundability of taxonomy, for Godfray, depends on the extent to which a definable and achievable product can be offered. For him, the Web offered an opportunity for a fundable product to be defined in the shape of "Web revisions": moderated taxonomic treatments of groups of organisms, which would become the new starting point for the taxonomy of the group. Godfray envisaged a refereeing process to agree on these new consensus taxonomies, which would be made available on the Web with comprehensive information and images. The idea of a consensus taxonomy is controversial to a field such as systematics which has a hard-won scientific status and so might be wary of any proposals that seem to reduce its standing to that of a bureaucratic mechanism. The notion that fundable products are needed for taxonomy to remain sustainable is, however, more widely shared. The proliferation of brand-named projects, many of which focus on a database or Web initiative, is in large part attributable to ideas of what is fundable in the current climate. A climate of doubts over the willingness of research councils to fund the ongoing practices of taxonomy, seen as the product of a mistaken focus on hypothesis-driven and innovative research, produces the perceived need to tap into alternative sources of funding by presenting sets of activities as discrete projects. Many of these new brands are the initiative of individual researchers and institutions. The two the Select Committee bought into, however, were brands that promised a far broader vision than an individual institution could achieve.

With respect to the use of ICTs, the Select Committee responded to the mass of evidence given to them with a straightforward map encompassing two brands, the Godfray vision and GBIF: "We recommend that the United Kingdom should take the lead and propose to the Global Biodiversity Information Facility (GBIF) that the GBIF run a pilot with some priority species to form the basis of a trial for Professor Godfray's suggestion of making taxonomy primarily digitized and web-based. A trial would demonstrate the benefits and pit-falls of this approach before implementing it more widely" (Select Committee on Science and Technology 2002a: 6). Guice (1999) describes the need for policy documents to present a future that is achievably close, and yet needs effort and investment to make real (otherwise, there would be no need for a policy intervention). Bloomfield and Vurdubakis (1994) discuss the sense in which IT strategy reports from consultants can be seen as textual constructions of reality that allocate responsibility between "social" and "technical" issues. Similarly, van Lente and Rip (1998: 223) describe a "dialectics of promise," in which scientists have to demonstrate their field's strength in terms of what it can offer while at the same time showing its weakness in that it needs funding and support in order to succeed and survive. The evidence to the Select Committee, and the final report itself, amply demonstrate these tensions. The choice of two brands to organize the recommendations on use of ICTs arguably presents a scenario carefully balanced between an existing project and a future vision.

In their commentaries on the government's response to the report, the systematics institutions make clear that the bundling of brands recommended by the Select Committee is not an entirely practical solution: GBIF is not a suitable place to try out the Godfray vision. While the responses are couched in practical terms, it also appears that the brands which the Select Committee found so compelling in the context of the report are not entirely helpful, to the systematics institutions, as a representation of the scope and complexity of existing activities. While the report focuses on recommending a few concepts or initiatives, the response from the systematics institutions is that they are doing as much as they can on a number of fronts, but that the situation is inherently complex. As Ogden, Walt, and Lush (2003) argue, a simplified branded policy might be disseminated successfully, but may risk overriding diverse yet locally appropriate existing activities. The response of the systematics institutions suggests a similar emotion. Orienting initiatives toward particular audiences and funding opportunities is also not without its tensions and worries about "selling out," as Cooper and Woolgar (1996) illustrate. Brands are therefore somewhat troubling concepts in science policy and science practice: although they can be compelling precisely for their breadth, because of the neatness of their packaged solutions and for the qualities they evoke, for these same reasons they can be viewed by participants as missing out on the reality of what they do and as diversions from the real business at hand.

A final brand that comes through as significant in the 2002 Select Committee report and is mentioned in the overall recommendations is the Darwin Initiative. This stream of funding is a direct response to the commitments made at the Rio summit, and demonstrates the UK's willingness to meet its responsibilities to share taxonomic expertise and information with less rich countries. As a brand, then, the Darwin Initiative within the report represents the UK as a responsible leader in a global field. The Select Committee endorse this branded initiative, both as a way of meeting global responsibilities and as a means to increase the digitization of collections. In practical terms, however, this initiative is going to be able to fund only a limited set of small-scale projects, on a budget of £3 million per annum from 1992 to 2002, not all of which is spent on taxonomic projects. Arguably, the branding of the Darwin Initiative, as well as the prominent position it occupies in the Select Committee report, presents a problem for systematics. The Darwin Initiative encompasses so many desirable qualities for relatively little money that it could make other means of addressing global inequalities in taxonomic expertise seem redundant.

To summarize the upshot of this theme, the use of ICTs in systematics as portrayed in the report appears to be a highly self-conscious practice. An awareness of the strategic significance of systematics, an understanding of the desirable qualities of the Internet as an evocative object, and a set of beliefs about what form of activity in systematics is likely to be fundable give rise to the focus on particular branded initiatives. This focus comes through in the evidence of the major systematics institutions to the Select

Committee. Each institution describes its own in-house initiatives, while also demonstrating its involvement in key international and transinstitutional activities. Strategic thinking is by no means new in science, but the prominence of branded initiatives in the case of systematics places a new importance on questions about the visibility of initiatives and new forms of organization in science. In chapter 6 I explore in more depth the resulting structuring of systematics in terms of institutions and initiatives. It appears from the highly selective set of data in the report that it may be important for projects that aim to attract attention and funding to consciously develop brand qualities that will be attractive. Developing the appropriate image can be seen as crucial to success in a competitive funding sphere. This, of course, will not be news to many working in science (and social science) today. The need for a snappy title or a catchy and evocative acronym has occupied many a project team along with more substantial issues of aims and content.

Historical Specificity

It is clear from the Select Committee's report that ICTs in systematics are entering a highly charged territory. Talking about these technologies, in the context of the 2002 Select Committee inquiry, involves co-constructions that shape the identities of technologies, people, and organizations in sets of moral relationships, conferring roles, rights, and responsibilities. In talk like this the stakes are high, and all of those involved appear to be conscious of the fact. Contributions are carefully crafted to make the best impression. Systematics institutions are bidding for the future of their institutions and their discipline, while government departments, funding bodies, and learned societies all have past decisions to defend and future priorities to protect. The status of this evidence as a reflection of practices of ICT use in systematics therefore needs to be understood with some caution. This is not a neutral portrayal of how systematics is using ICTs. The themes that have emerged in the analysis need to be followed through in other arenas in order to assess their wider currency.

Four thematic areas emerge from the report. The first, focusing on the relationship between physical specimens and virtual information, demonstrates that the capacities of technologies to achieve particular functions for taxonomy are contingent, open to debate and interpretation. The second theme, the complex political geographies of taxonomy, shows that the adequacy of virtual taxonomy is played out on a global stage, in which local initiatives can be influenced by and consequential for international relations of responsibility and dependency. The third theme shows that representations of what the technology can and cannot do are tied up with representations of the need for and qualities of human expertise, and that such representations again take place in a highly charged territory where funding decisions can depend on appearing to be suitably innovative and yet not replaceable by innovative use of technology. Finally, the fourth theme demonstrates the extent to which particular technologies become appealing thanks to the qualities they evoke, rather than any practical functions they may perform. This evocation is sustained by reference to the qualities the technologies have for other communities, filtered through particular sets of local concerns and priorities. The focus on evoked qualities also applies to individual projects, initiatives, and visions, where the development of an appropriate brand image is important for successfully enrolling policy makers and funders.

In other parts of this research I examine the use of mailing lists by taxonomists, both in terms of the extent to which the themes described above arise in discussions and as an alternative context in which taxonomists engage in reflection on their discipline. I also undertake research into the evolving landscape of institutions and into the digitization of specimen collections. By looking at ICTs in systematics across these different contexts we can see that, as the themes described above suggest, ICTs are embedded into systematics in very specific ways. Far from a unidirectional transformation of a discipline and its working practices, we find culturally specific embedding of technologies into particular contexts. At the same time, however, ICTs provide a cultural field for the playing out of systematics, and this field enables both the development of new practices and the opportunity for reflection on these practices.

The Select Committee report is, of course, a highly unusual source of data and a particularly troubled one to offer up as a representation of systematics as it actually is. Evidence was presented by systematics institutions acutely aware that their prospects for future funding depended on making a compelling case. Funding bodies and government departments gave evidence to the Select Committee conscious of the need to justify past decisions and protect their own complex agendas. The themes that I have identified arise within this very particular context. They also arise at a time when developments in global politics have given biodiversity information, and hence systematics, an unprecedented profile. The coming of the Internet also plays a strong part in the development of the themes, since the possibilities of the Internet are highly influential in imagining the future of systematics. This raises a question as to whether the themes identified here are entirely of the moment, or whether they would have arisen in similar fashion in the past. Are these themes—physical and virtual specimens, complex political geographies, automation and expertise, branding and evocative objects—entirely themes of early twenty-first century systematics, or is this merely a new incarnation for old concerns?

This question cannot, of course, be answered unequivocally. A comparable set of evidence for the systematics of previous periods is just not available. It is, however, possible to look at some selected times when use of information technology in systematics has been a matter for explicit promotion or debate and use these as some index to the development of themes of concern over time. The potential of computers for taxonomic work has been recognized since at least the early 1960s (Hagen 2001). An international conference on Systematic Biology in 1967 included a session on "Computer Techniques in Systematics," and one discussant at that session claimed twenty years of experience with computers. A brief description of this conference session should serve to show what some of the key concerns of that time were, before we move more systematically to consider the currency of the particular themes explored in this chapter.

Although the demonstration of using a teletype system to interrogate a remote identification key may seem rather quaint in format now, much of the 1967 discussion still seems fresh, at least in the kinds of issue raised. The session began with the author attempting to debunk some misconceptions about computing: "It almost seems, from some of the exaggeration in the public press, that if one should trip over a power cord while walking by a computer it would be stimulated to spew forth a revision of the Carabidae" (Bossert 1969: 595). The paper goes on from this point to stress that the computer, rather than replacing the taxonomist, provides a tool that needs to be used with care and with comprehension of the underlying algorithms. At this stage, just as in the evidence to the 2002 Select Committee, care is taken to point out that computers do not replace taxonomists, although here the warning is framed as much to taxonomists themselves as to outsiders. The main concerns then discussed are rather less familiar, being located in the context of the hardware available at the time: Bossert spends considerable time discussing how time-sharing computers might make computing power readily available to taxonomists in their work. Availability of computing power is seen as the limiting step in application of computers not just to statistical calculations in numerical taxonomy, but to a broader role in storage and interrogation of taxonomic information. Bossert then goes on to map out a vision of a centralized store of taxonomic information, of a form which would be thoroughly familiar to the authors of the 2002 Select Committee report and the participants in the GBIF program. Bossert considered this would be "technically or economically feasible by the early or middle 1970s" (Bossert 1969: 603). The main problems that he foresaw were overcoming conflicts between the use of a centralized information store and the traditional publishing system, and getting computer scientists and taxonomists to work together to realize the system. Discussants endorse Bossert's vision of a centralized taxonomic information store, adding the potential of computers to take on routine tasks of making inventories and tracking data, to scan and store images, to extend capacity for exploring numerically based classifications, and to allow for communication and sharing of expertise between institutions in real time. Hardware and software may have changed radically, but the visions of what computers might do for taxonomy have remained remarkably unchanged.

Visions of wholesale change to the systems of research and communication in systematics are certainly not new. To illustrate, I quote at length here a piece from the published version of a 1969 symposium:

> The Flora of the future will be a standardized data bank from which all "cutting and pasting" can be done. It will be open-ended, dynamic, and ever growing. If such a bank existed today, then Flora North America could be written overnight by machine, except for the taxonomic judgements required. The specialist of the future doubtless will store his latest revisions and monographs in the computer, not on the printed page. He will deposit his data in a computer that is available nationally if not internationally by a telecommunications network. Such networks already are in operation. Thus the Flora of the future will become a huge memory or series of linked memories available on-line to all users at any place and time.
>
> Computers can and will affect the very process of research and communication in floristics and taxonomy. . . . The users themselves will be able to dictate the parameters and format of retrieval output. The scientific journals and books can spare their pages of endless descriptive facts and devote them exclusively to scholarly, theoretical discourse. (Shetler et al. 1971: 306)

Clearly, visions of "Web-based systematics" as articulated by Godfray (2002a) are far from new. Shetler and colleagues, based on their experiences of constructing the Flora North America, were prepared to make radical predictions about what the discipline might become, although they stopped short of the contemporary suggestion that a consensus taxonomy should be produced. Shetler and colleagues saw the centralized databank as a direct conduit from systematists to users without any mediating steps. Visions of user needs have therefore changed over time. However, it is apparent that many factors we might recognize from the current preoccupations were visible in the concerns of the discipline in the late 1960s, though different issues have taken more or less prominence over the course of time. I have demonstrated elsewhere (Hine 1995) that interest in the role of information technology in systematics has taken different forms. Here I revisit that analysis in light of the themes drawn from the 2002 Select Committee report. Previously I argued that from the 1970s to the 1990s, three separate sets of concerns in systematics relevant to the use of information technology can be identified. In each case, the particular set of problems and the information technology solutions proposed were quite different, but in each case the concern was with the status of systematics as a discipline.

The first area of concern with the use of information technology in systematics revolved around the methods used to group organisms and identify groupings. Two key positions, phenetics and cladistics, came into prominence in the 1960s and 1970s and formed a focus for disciplinary splits and tensions: the dynamics of the development of these positions and the ensuing debates are described in lively style by Hull (1988). Each approach was opposed to the other, and to the traditional "craft" practices of taxonomy. Phenetics uses numerical methods to group organisms on the basis of overall similarity, whereas cladistics works on the basis of inferring evolutionary descent. In both cases, the need to make taxonomic methods explicit and objective was key, and both methods drew on computerized analysis. Pheneticists aimed to develop algorithms that would group organisms based on their overall similarity, removing the responsibility from the taxonomist to decide which organisms were most like one another. Computers were essential to perform the necessary calculations. Cladistics also used computers to perform analyses, developing methods for unpacking complex patterns of shared characteristics into inferences of evolutionary descent. The computational aspect was not so explicit in cladistics, but in practice carrying out a cladistic analysis rapidly came to mean using a computer package. The relative merits of phenetics and cladistics were hotly debated, including some high-profile disputes in *Nature*. Each claimed adherence to scientific method and stressed the objective nature of the classifications that resulted. The role of computers in assisting the analysis was useful in this rhetorical context, in stressing that it was not the taxonomist who made decisions, but the computational algorithm.

Computerized support for taxonomic analysis therefore figures as an important feature in both cladistics and phenetics. Each approach was presented as a response to the needs of the discipline for more repeatable methods and hence improved scientific standing. As Hagen (2001: 308) puts it, "claims for objectivity were a staple in the literature of computers in systematics." A similar set of issues can be found in a very different arena in concerns over the objectivity of medical decision-making as described by Berg (1997). Here lack of consistency between medical experts acted as some justification for the introduction of decision-support technologies and expert systems. Just as traditional taxonomy could be portrayed as subject to the whims of an individual taxonomist, so too could traditional medicine be depicted as varying with the particular physicians concerned. In both cases, this variation is viewed as detrimental to the image of the profession. The availability of information technology in each case could be said both to offer solutions to the perceived problem and to contribute to the articulation of the problem in the first place.

Hagen (2001) also suggests that the availability of computing and the broader cultural appeal of computers as replacements for the human brain were influential in shaping the course of systematics. He suggests that the concept of parsimony in particular became widely discussed because it was amenable to formalization rather than because it had appeal on theoretical grounds. In Hagen's analysis of the introduction of computers into systematics in the 1960s, then, we find some very familiar themes. Systematists of that era are worried about the status of their discipline, keen to appropriate some of the cachet and air of objectivity that computers bring with them, and drawn in particular directions by notions of computable problems. For Hagen, the turn toward computing was prompted by the availability of large data sets. Whereas later on it appeared that the debate had always occurred along sharply drawn lines between cladistics and phenetics, for Hagen the distinction was not so clearly drawn at the outset, and many of these beliefs about computing were shared by both sides.

The cladistics–phenetics debate was settled largely in favor of cladistics, the weight of opinion having come to rest on classifications as reflective of evolutionary descent rather than attempting to reflect overall similarity in phenetic fashion. Debates still continue about the precise relationship between cladistics as a means of inferring evolutionary relationships and the practicalities of delimiting and naming a taxon (unit of classification such as species, genus, family). However, cladistic analysis itself has largely become established, and within that context the use of computerized analysis is viewed as unproblematic. This position is reflected in the evidence given to the Select Committee in 2002. Little examination is given in that report to the scientific status of taxonomic method, or indeed to any of the details of how taxonomists assign organisms to taxa. To this extent, then, a previously prominent theme in the use of ICTs in systematics—that of scientific method and objective knowledge—seems to have receded in recent years, or at least does not merit consideration in the public arena of the Select Committee inquiry. Where concerns over method are still alive, they revolve around nomenclatural systems.

The second area of historical interest in uses of information technology in systematics focuses on automation and alleviating the burden of routine work. The interest in computers was as a means of saving the time of taxonomists for expert work which only they could carry out. Generalized databases containing coded descriptions were to be created, which would then enable identification keys, identification programs, and descriptive monographs to be produced automatically. Databases were to be used to maintain catalogs of specimens and assist in their curation. The focus in each case was on protecting the valuable time of the taxonomist and curator. Themes of automation and expertise identified in the data of the 2002 Select Committee report have therefore been alive in the discipline for some time. Many of the concerns about demarcating expert work from tasks open to automation that were raised by systematics institutions in 2002 can be glimpsed in the discussions at Systematics Association symposia in 1973 and 1984 (Pankhurst 1973; Allkin and Bisby 1984).

A third area of concern where information technology became relevant to systematics was that of meeting user needs for biological information and stable nomenclature. As with the cladistics–phenetics debate, a series of letters to *Nature* in the late 1980s and early 1990s marked a public (among biologists at least) concern. Rather than caring about the scientific status or objectivity of taxonomic methods, ordinary biologists were presented as interested in stable sets of names. Many of the responses to this perceived problem revolved around proposing changes to the nomenclatural codes that legislate how names are assigned to taxa. The dispute also fueled a movement focused on improving the image of taxonomy by providing products and information services directed at making information accessible to users in forms they could understand. Elements of this position can be glimpsed in earlier discussions of the potential of computers to produce outputs in forms dictated by user preferences (Pankhurst and Walters 1971; Shetler et al. 1971). User needs became more prominent in discussions in 1984 (Allkin and Bisby 1984), and gained a new visibility and a new urgency by the early 1990s: "We believe systematics can be saved from its present crisis by the development of user-oriented information products. The modern world values information services highly and is willing to pay for them. Taxonomy is a biological information system from which valuable products and services could be produced to meet the demands from the much expanded interest in all aspects of species diversity" (Bisby and Hawksworth 1991: 332). Here we see systematics itself being reconceptualized as an information system. Information technology, as an evocative object, provides a model for thinking about what systematics does, which suggests particular visions of future action.

It appears, then, that the themes identified in the Select Committee report of 2002 have some relevance as ways of understanding the history of ICTs in systematics, although the balance has shifted with recent concerns, and some previously prominent issues have receded. In particular, the concern with portraying systematics as a scientifically valid practice has receded into the background of public debate as cladistic analysis has become a taken-for-granted methodology and as the use value of systematics in terms of biodiversity conservation has become more prominent. The theme of automation and expertise has been a continuing concern since the early history of information technology in systematics. The earlier interest in automating routine tasks and preserving the time of taxonomists for jobs requiring real expertise makes the same kind of demarcations that we see in operation in the 2002 Select Committee report. Just as before, we find taxonomists concerned to point out that routine tasks have to be done, that they cost money to do, and that they detract from doing the work that they and their users feel they should be focusing on. Concerns with users persist, but have often been translated into the domain of international politics and commitments to biodiversity conservation. Interests in the qualities of virtual and material specimens and the audiences for availability have been articulated in increasing detail, thanks in large part to the increased confidence that usable taxonomic information systems can be achieved with currently available technologies. The seeds of this interest can be seen in the earlier debates on biological identification with computers and the possibility of a general-purpose descriptive database. Thus the picture of systematics mapped out in the Select Committee enquiry of 2002 has firm roots in the development of the discipline and its relationship with ICTs over the last thirty years, although emphases have shifted and the specification of concepts has altered.

Another way in which to judge the historically located nature of the 2002 report is to look at its predecessor, the 1992 Select Committee "Systematic Biology Research" report (Select Committee on Science and Technology 1991). The Committee of 1991, chaired by Lord Dainton, conducted a wide-ranging inquiry into the state of systematics in the UK, prompted by protests about a funding crisis at the Natural History Museum in London that was felt to be threatening its ability to curate collections effectively and carry out taxonomic research. As with the 2002 report, it is not appropriate here to consider the full scope of the report and the arguments advanced in evidence. Instead I will look only at the sections of the report that discuss use of information technology, for the sake of comparison with the 2002 report.

Information technology did feature explicitly in the Select Committee's call for evidence. Among ten questions posed by the invitation to submit written evidence was the following: "Is the availability of information technology (computerized databases) to systematic research being adequately exploited? Is United Kingdom research taking cognizance of the full range of new developments in this field?" (Select Committee on Science and Technology 1991: 99). Clearly, then, the report positions information technology as a coming technology in systematics, and one that offers some (unspecified) promise. In the main body of the report, however, the role of ICTs is very low key compared to the 2002 report.

In 2002 the need for full specimen catalogs to be available on the Internet is taken almost as given, whereas in 1991 the chapter of the report focusing on curation makes no mention of information technology. Information technology is discussed at length only in a chapter on "modern methods" in systematics, and even here any radical significance for the discipline is denied: "Within the last decade only, three additional techniques have made their impact and perhaps qualify as "new" tools in biological systematics—molecular biology (in the form of "molecular systematics"), information technology and image analysis. While the first of these has generated an entirely new category of systematic information, the latter two are essentially merely ways of handling data more swiftly or productively" (Select Committee on Science and Technology 1991: 46). The report enumerates three ways in which information technology is thought to be of use in systematics: in providing specimen inventories, allowing specimens to be located easily and labels generated; in speedy analysis of large amounts of data, such as needed for cladistic analysis; and as a means to automating identification and access to information for non-experts. The association in each of these instances is between information technology and efficiency, removing laborious work, speeding up analysis, and removing the need for expert involvement in identification. The report then goes on to enumerate various initiatives to develop taxonomic databases, considering them "patchy" and possible lagging behind efforts in the United States. Although the seeds of many of the current initiatives can be seen in the cases that the report lists, it is striking that the 1991 version makes little if any reference to problems caused by lack of access to taxonomic information. Some individual pieces of evidence do discuss concerns: the Natural History Museum notes the problems of cataloging specimens, CAB International makes reference to developing international standards, and Dr. Frank Bisby (leader of the ILDIS project which I discuss in chapter 6) gives evidence at some length on the potential of taxonomic databases and the concerns over their funding within current structures. Significantly, few of these concerns are featured in the main report, and where they are considered they are given low priority.

It is clear that although the Select Committee of 1991 saw that information technology could be used in systematics, they largely saw its significance as a means of making work more efficient or more objective, in line with the debates in systematics of the 1970s and 1980s, and failed to adopt the view of systematics as an information system focusing on user needs. Information technology figured as desirable in that its use was viewed with approval and it was seen as a bad thing to be lagging behind other countries. There was, however, no rhetoric of universal data availability or accessibility to illuminate this vision. Information technology was seen as a means of access to information but largely within a closed taxonomic community, and the virtual specimen had yet to come into being as a credible object of discussion. The political geography that informed the report was one of concern over the colonial heritage of British systematics, seen largely as conferring a funding burden on the UK along with the moral responsibility to maintain collections. In a stark contrast to the discussion of data repatriation in the 2002 report, the 1991 committee gave serious consideration to the prospect of repatriating the specimens themselves in order to save money. Political and geographic sensibilities were different then, as were attitudes toward public spending, and the technologies that shaped expectations were different too. Thus seeds of the concerns of 2002 are visible in the 1991 report, although in many respects the two are so far apart as to represent different worlds.

Evocative objects prove key to understanding much of the shift in emphasis and respecification of themes in the recent history of systematics and ICTs. The available technologies provide different ways of thinking through the possibilities and problems of systematics, and at each stage in its history systematics imagines itself through the available technologies. These processes of imagining both draw on and feed into the wider cultural currents of the time. At each stage, however, these imaginings have to fit in with complex political geographies and funding environments, themselves swayed by their own sets of evocative objects. The environment in which this occurs is provided by the positioning of systematics as a strategic science, in the multiple senses of that term described by van Lente and Rip

(1998). Systematics has been increasingly seen as a strategic resource in the conservation of biodiversity, and to this end has become embedded in the international politics of this effort. Through this increased prominence the activities of systematics have become the object of strategic policy concern and intervention. Within systematics, the activities of practitioners, funders, policy makers, and users have consequently become imbued with strategic significance, and many have come to see their actions as in need of strategic thinking. This is consistent with what, for van Lente and Rip, is an intrinsic part of contemporary science: "Present-day scientific-technical communities are strategically oriented. They are creating a kind of science where promises and assessments, and the rhetorics (and counter-rhetorics) that go with them, are an integral part of 'doing science'. (van Lente and Rip 1998: 245). In multiple senses and for both practitioners and policy makers, the imaginings of systematics and the uses to which it puts information technology have become thoroughly strategic.

Derrida (1998) wondered whether Freud's understanding of memory would have been fundamentally different, had he had different technologies through which to imagine it. We can ask in a similar fashion whether systematics, or indeed any science, might be very different if it had different technologies through which to imagine itself. The technologies that science uses are functional, in that each scientific instrument, as we saw in the previous chapter, makes particular features of nature visible and manipulable. These same technologies are, however, developed according to visions of possibility and necessity, ideas of what could and should be done. The history of ICTs in systematics is, as this chapter has shown, subject to visions of what the discipline is and what it should become, which are in turn intrinsically linked to the technologies available for carrying out and imagining that work. As we see in the work of the Select Committee, the interested parties imagining future systematics include the systematists themselves, both as individuals and institutions, but also include a cast of nongovernmental organizations, government departments, and funding bodies who all have a stake in imagining the discipline as well. The Select Committee inquiry provides a very public forum for this imaginative work, where the stakes are high. As van Lente and Rip (1998) describe it, this kind of talk about the promises of science and technology creates a rhetorical space, ripe for filling in with social reality as resources are mobilized and networks formed. The rest of this book looks at the social realities of ICTs in systematics as they have developed over time, while always bearing in mind that the distinction between rhetorics and realities is a troubled one, and that the social reality is always understood through some form of rhetoric.

4 Behind the Scenes and Across the Globe: Virtualizing the Material Culture of Systematics

As material cultures, the sciences give meaning to objects and embed them in working practices. Technologies, laboratory animals, and other experimental objects become part of a material culture which defines whom they belong to, how they should be acquired, stored, and disposed of, and what features are relevant to working practice and to the communication of scientific results. In turn, the distinctive material cultures of scientific disciplines combine to provide ways of manipulating and making visible the natural world through experimental systems and scientific instruments. Given this emphasis on material culture, and given that different branches of science may differ quite markedly in their understanding of material objects, it is appropriate to explore how systematics is organized in this regard in order to develop a picture of the territory in which ICTs intervene. An interest in systematics as virtual culture needs to be situated in a thorough grasp of the material culture with which that growing virtual culture is enmeshed. In the previous chapter we saw particular qualities being attributed to virtual objects in a policy document, notably that they were an accessible alternative to material specimens. In this chapter, then, I set out to explore the material cultures of systematics on the ground to find out how virtual objects play out in practice. The treatment is thematic. Within each theme I explore a significant aspect of the material culture of systematics and consider the ways in which initiatives in use of databases and computer-mediated communication are fitting into and transforming that aspect of culture.

The five main themes treated in this chapter are: specimen collections as scientific instruments; images and working practices; organizing digitization; the ordering of objects; and systematics on display. These five themes reflect important aspects of the material culture of the field, and also allow a discussion of some of the qualities that ICTs have been able to bring to the field. In the first instance, looking at the role of specimen collections as scientific instruments in systematics allows me to explore how databases of specimen collections have come into being and the priorities that have dominated their design and implementation. The second theme looks in more detail at the ways in which digital artifacts have been incorporated into working practices, and the third theme then looks at the working practices and organizational arrangements that give rise to digital objects. In the fourth theme, on the ordering of objects, we find that systematics has some cherished ways of arranging objects, both for pedagogic purposes and for their integration into the practices of systematics. The notorious ability of databases to provide improved and flexible access is an additional stimulus to the computerized cataloging of collections. The ways in which physical and virtual arrangements of objects supplement one another are explored in some depth. In the final theme, the status of systematics as a culture of display is considered. The changing politics of publicly funded science and of museums provide an impetus toward increased accessibility for systematics collections. In this context too, a virtual culture of display has developed and is interwoven with the display of material collections.

The cumulative effect of these five themes is to demonstrate that systematics is a vibrant material culture, which has strong roots in established tradition and historical practices, but yet is experiencing change in response to new techniques, new technologies, and new political pressures. As a part of this, a virtual culture is emerging that is closely entwined with the material culture. Neither is a fixed point, but each provides for the mutual elaboration of the other. Virtual objects have provided fruitful sites for imagining a future for specimen collections, but this vision is not one of wholesale transformation. The imagining of change in relation to the material culture of systematics has strong links to the heritage of the discipline.

Specimen Collections as Scientific Instruments

The first point to be made is just how intensely material a culture systematics is. The discipline founds itself on the study of the diversity of living things, which is made manifest in collections of specimens. Some sense of this wealth of material culture, and the importance placed upon it, can be seen in a recent article on botanic gardens by the Secretary General of Botanic Gardens Conservation International:

> Their collections of living plants, herbaria, libraries, seedbanks, works of art, manuscript archives and others contain tens of millions of specimens, representing an unequalled library of life and a window to the plant kingdom. Botanic garden herbaria alone contain almost 150 million preserved plant specimens, collections on which the

scientific names and descriptions of most plant species have been based. Together, botanic gardens grow up to 100,000 different plant species—perhaps a quarter of the world's plants, represented by some 6 million individual plant holdings. What Noah did for animals, botanic gardens are doing for plants. (Wyse Jackson 2003)

The article continues by recounting the recent growth in numbers of botanic gardens, situating growth of interest in living collections of plants in the increasing awareness of biodiversity conservation. While this new emphasis places fresh importance on the garden as a bank of material for understanding ecosystems and researching conservation and sustainable plant use, the heritage of gardens is also, as Wyse Jackson (2003) describes it, as a "library of life." As chapter 3 showed, concerns with biodiversity conservation and systematics are thoroughly entwined in the current political and funding climate. Nonetheless, as far as systematics is concerned, the role of collections of specimens is defined independently of the politics of biodiversity. They are seen as providing a "library" which the systematist can consult. In the description of specimen collections as a library of nature there are some clear resonances with views of the laboratory developed in sociology of scientific knowledge. Just as the laboratory can be seen as a place where natural phenomena are brought under control and made amenable to scientific study, so the natural history collection brings together the variability of organisms on a scale the systematist can survey.

The working practices of systematics depend on the accessibility of example organisms that can be studied, juxtaposed, and compared—a point that might be assumed in the daily practices of taxonomists but is made explicit in introductory taxonomic texts (e.g., Heywood 1976; Stace 1980, 1991; Jeffrey 1982). The foundational nature of herbaria for systematic botany, in research and training and in the provision of a focus for work in organismal biology, also tends to be stated explicitly when funding restrictions place collections under threat, as occurred recently at the University of Iowa Herbarium (Horton 2003). The American Association of Museums has issued a statement decrying the financial threat to collections and outlining their vital status: "At risk are collections of irreplaceable objects, such as geological, paleontological, zoological and botanical specimens, anthropological and historical artifacts, and archives. These collections are held in trust for the public; they are the priceless heritage of this and future generations; and they constitute critically important resources for new knowledge" (American Association of Museums 2003). The positioning of collections as a resource, and specifically as a reference resource to be consulted in future (often undefined) projects, is a common feature in descriptions of systematics. Wyse Jackson (2003), we've seen, describes botanic collections as a library; on behind-the-scenes tours of the Darwin Centre at the Natural History Museum (store of the museum's collection of animal specimens preserved in spirit) and of the herbarium at the Royal Botanic Gardens, Kew, I was fascinated at both sites to hear the collection described as a database. The connotations of information storage and retrieval now familiar from computer technology have been appropriated as a means of understanding the role of collections. The metaphor of the database works, for practitioners, without any actual computing technology needing to be involved. However, along with the metaphorical use of the database as a way of understanding collections of specimens, the actual use of databases as ways of cataloging collections and communicating systematics has become increasingly important. I will demonstrate that this expanding virtual culture of systematics fits into, transforms, and strengthens the material culture, providing a lens for imagining change and stressing continuity. To understand how this happens it is necessary to explore some features of the material culture in more depth, addressing the ways in which systematics currently construes its collections of specimens as the right tools for the job (Clarke and Fujimura 1992).

Within systematics, specimen collections take a wide variety of forms, including: the living collections of botanic gardens, culture collections of fungi and microorganisms, and, to a lesser extent, zoos and aquaria; herbarium specimens, largely dried and pressed on card; seed banks; animal skins or skeletons; insects preserved whole, usually pinned in drawers; spirit collections of animals or plants, largely in glass jars; fossils; DNA or tissue banks; slide collections of prepared materials; and collections of bulky specimens, such as wood. The value of specimens lies in their ability to stand in for living organisms in their natural habitats, for the classifications based on them are intended as an index not merely to the collection, but also to the diversity of organisms in their usual living state. In each case, the specimen is preserved in a way that is intended to keep as many as possible of the features that systematists will want to study in future, and to be as long-lasting as possible. A dried fungus specimen might look uninspiring, but for fungal systematists many of the microscopic characteristics important for identification remain unaffected by drying. An herbarium specimen can last for centuries in a usable form, provided it is protected from humidity and insect damage (a feat more easily achieved in temperate climates). Many of the morphological characteristics that would interest a systematist are available in a pressed plant specimen: and of course, over time systematists will be drawn to studying precisely those characteristics that have been preserved in the dried form.

There is bound to be some circularity between the characteristics that are valued for systematic purposes and those that happen to be available in preserved specimens. Color is unpopular as a basis for systematic work, being both hard to preserve, whether in dried specimens or spirit preservation, and also hard to characterize objectively. I will tend in this chapter not to dwell on fossil specimens, discussing only those organisms that are collected in the live state. Fossils, however, do raise special concerns for taxonomists thanks to the nature of their preservation. In any given fossil specimen only small parts of an organism may be present, and these may be preserved in ways that do not make many useful characteristics available for taxonomic purposes. Bowker (2000) discusses the problems this creates for a stable taxonomic treatment of fossils, such that different forms of preservation may lead to different taxa being identified, and, more fundamentally, different criteria being applied as to what constitutes a taxon. The means of preservation and the resulting taxonomies are not independent, a concern that certainly some taxonomists dealing with some groups of organisms recognize. This, for Bowker, raises a considerable problem for biodiversity databases, which may inadvertently solidify such distortions and make them more widely available to be taken as a true representation of the way the world naturally is. Bowker's picture of the taxonomic system as an instrument for viewing natural diversity is that it is intrinsically flawed, given our inability ever to produce systems of naming and classification that are free from being influenced by the vagaries of the material available and the practices of taxonomists. His view of databases is that they can make things seem more real than they should be. This is a theme I will be developing further in this chapter and which we will see is a fear shared also by taxonomists.

Specimen preparation and curation is a technical matter, which requires experience and training. There are numerous guides available to assist collectors and curators in selecting, preparing, and storing specimens to best advantage (recent examples include Bridson and Forman 1998 and Metsger and Byers 1999). The Royal Botanic Gardens at Kew organizes regular eight-week courses on herbarium curation technique. The means of preparing specimens, including the glues or straps used for attaching specimen to paper, the type and location of label, and the methods for protecting against pest damage, will vary between institutions, but in each case constitute thoroughly researched and established practices rather than matters of chance or individual choice. Warnings about using the wrong kind of ink or label can be fierce, and it is commonplace for visitors to collections to be warned to follow local practice carefully and ask local curators for advice.

Specimens of plants are routinely pressed and glued to sheets of paper. This means of preserving plant specimens in collections is attributed to the sixteenth-century Italian botanist Luca Ghini (Pedrotti 1995). Linnaeus, working in the eighteenth century, is credited with the innovation of leaving the sheets unbound, allowing specimens to be moved for ease of visual comparison and reordered into classification schemes (Pedrotti 1995). Some older sheets also have specimens of more than one species mounted on the same sheet: a practice that is certainly not acceptable in the preparation of contemporary specimens. The acceptable means of preserving specimens thus depends on ideas of how they will be worked with and how they will be stored. Herbarium specimens of the modern single-sheet variety are typically kept in cardboard folders, stacked in cupboards or compactor units. In the tropics there may be need for a controlled climate, but this is less often the case in more temperate regions. Perceptions of the best way to store specimens have changed over time. The mycology herbarium at the Royal Botanic Gardens, Kew, possesses Victorian specimens which, the fungi having been viewed at that time as primitive plants, were mounted as plants on herbarium sheets. This practice would often be a struggle for bulky specimens, and might require a specimen to be sliced up. Modern practices view the fungi as worthy of classification and curation in their own right, and keep specimens in paper packets in file boxes in a form that better respects the shape and size of the whole organisms.

Specimens are typically stored in ways that suit the physical characteristics of the preserved items as they are understood to be important within contemporary systematic practice. A description of the way in which the collection of wood specimens in the Padua Botanic Garden collection is stored gives some insight into the specificity of storage arrangements for a particular kind of specimen, the care with which arrangements for storage are made, and the respect now given to the detail of these arrangements. "Each specimen—a complete section several centimeters thick—is now contained in a heavy, glossy paper bag carrying a label applied with gum Arabic and bearing the words HERBARIUM HORTI BOTANICI PATAVINI, the typewritten position number according to the Dalla Torre and Harms catalogue, and the taxon of origin. Closed with a pin, the bags are stored in wooden boxes lined with white paper, whereas the outside is covered with fabric and marbled paper" (Giulini 1995: 260). Giulini also notes in some detail the qualities of paper, typewriting, handwriting, and form of taxonomic names, all of which give insight into the dating of this collection. Information on the origins of historical specimen collections is often incomplete. All labels and accompanying information available are kept as carefully as the specimen itself, as a potential future source of valuable data for interpretation of the specimen.

Spirit collections, comprising jars of specimens in an alcoholic medium that prevents decay and preserves three-dimensional structure (although again, not color), are notoriously bulky and also pose a considerable fire risk. In the 1990s the Natural History Museum in London, recognizing that its arrangements for storage of the spirit collection on open shelves in a building from the 1920s were considered inadequate, launched a fund-raising campaign that resulted in the building of the Darwin Centre. This project was to provide new facilities for storage, for scientific work, and for behind-the-scenes access to the public. The resulting building, opened in 2001, contains dedicated storage facilities for the 22 million specimens of the spirit collection, kept in compactor units at 13°C to minimize the risk of fire, in separately isolatable sections of the building. The first phase of the building was said to contain some 27 kilometres of shelving.

The Darwin Centre illustrates a common feature of natural history collections: the best conditions for storing specimens may not be also the best conditions for those working with them. The 13°C environment of the specimen stores strikes chill to the laboratory-coat-clad visitor. In the Darwin Centre, working space has been deliberately separated from storage space. In the ranks of compactor units that line the storage side of the building, there is nowhere to pause, to lay down a specimen for examination, to make a note or draw a sketch. Instead, the working space is all located on the other side of the building, and specimens must be collected and transferred there for examination. In the herbarium of the Royal Botanic Gardens, Kew, the working space is interspersed between the storage cupboards, and staff lay out their work-in-progress on benches. In previous times this arrangement might have been inadvisable, as posing significant dangers to staff working with collections; insect control, for example, was historically achieved by mercuric chloride, to the benefit of specimen preservation but at some considerable risk to the health of staff. Views on health and safety at work having shifted and new technologies having become available, the use of mercuric chloride is no longer permitted. Specimens are now routinely frozen for some days on arrival at museums or herbaria, in order to kill any pests present. Nonetheless, many specimens remain contaminated, and special measures (ventilation, wearing of gloves, opening of cupboards to allow vapor to disperse before removing specimens) are sometimes advised (Metsger and Byers 1999; Rader and Ison n.d.).

Various practical concerns therefore shape the way that specimens are stored, and consequently also shape the access that working taxonomists have to them. A typical working arrangement would involve a taxonomist bringing together a set of specimens of interest, either in a working space within the collection or outside it, together with a selection of relevant literature. Specimen collections often contain bookshelves, or collections of books, along with specimens, as it is routinely expected that the taxonomist will alternate between use of literature and examination of specimens. In the herbarium at the Royal Botanic Gardens, Kew, the cupboards are arranged in bays, each of which contains a desk, many of which are piled high with specimen sheets and books relating to the work in progress. Some work, or almost all for a microbial taxonomist, may be done in a laboratory environment. Increasingly a computer will now also be part of the working context, sometimes a taxonomist's own laptop brought into a collection, and sometimes a networked computer provided as part of the infrastructure of the collection or laboratory. Collection storage both shapes working practices, by meeting the demands of specimen preservation first, and is shaped by working practices, in that some means has to be provided for the taxonomist to bring together the various artifacts and sources of information that they need.

So much, then, for the storage of specimens and the ways of working with them. What, however, counts as a specimen, and where do they come from? Although the means by which specimens are collected from the wild is certainly sociologically interesting as a part of the material culture of the discipline, this aspect of the work of systematics is not entirely relevant to the aim of this chapter in discussing material culture as a means to understanding virtual culture. The process of creating a specimen will therefore be merely sketched here as it relates to working practices. A specimen is part of a plant, an animal, or indeed any living organism, that has been collected from its usual habitat and fixed in some way to make it available for systematic study. The display at the Darwin Centre explaining the basis of collections stresses that a specimen is more than just a piece of plant or animal; crucially, it has to be accompanied by information on its provenance to be a scientifically useful object.

Specimens arrive at museums and herbaria by a variety of means: they may be collected on expeditions organized by the institution that holds the collection, in the interests of pursuing a particular research project or filling a gap in existing holdings; they may be sent for identification or donated, either by another institution or by members of the public; or they may be acquired as whole collections, by purchase or bequest from other collectors. In each case, the process by which a living organism (or a piece of one) becomes a specimen within a collection is different. In each case, however, the process is both an intricate practical one of preserving and presenting the specimen and a theoretically driven one, in that these specimens are selected and preserved for the sake of giving some insight into the variation and distribution of living organisms. Collectors are warned to choose specimens that reflect the population of organisms best, rather than simply choosing those that best fit the collecting equipment (Massey 1974a).

Collecting gives systematics its hero stories. The history of the discipline contains many tales of the great collectors who amassed vast collections, and of the adventurers who underwent great hardship to bring back specimens from the least accessible parts of the world (Musgrave 1998; Rice 1999). This picture of collecting as adventuring is not confined to the distant past. An account of present-day plant collecting describes botanists as undergoing a "six hour adventure in a dug-out canoe on the trail of a mystery palm" (Gunn 2003: 24). Collecting is often portrayed in these popular venues as entailing hardship, adventure, and sacrifice, often in challenging tropical climates. Collecting in the more temperate regions can also be challenging, but here the rigors tend to involve being cold and wet, and the stories are less amenable to telling as heroic tales. Although collecting, as an active pursuit, is often portrayed as the core of collections, and is the key way of expanding a collection in particular directions to cover gaps or explore priority groups, in practice the material is often simply donated, whether by public or by other institutions. According to established convention, a botanist collecting a plant specimen which she deems important for her own institution may also deposit specimens in the national herbarium of the plant's country of origin, and may send out specimens to other institutions. This practice gains meaning through current sensitivities about the stripping away of natural heritage from its country of origin, and through the fragility of specimens, which requires multiples for insurance purposes and also to restrict the need to send specimens on loan.

The net result of collecting and donations is collections that can be awe-striking in scale and often in sheer bulk as well (the Natural History Museum in London, for example, is reported to have some 68 million items in its collections; McGirr 2000). Having amassed these collections, what, then, do systematists do with them? The answer, for very many of the specimens, is nothing at all for most of the time. These collections are archival, in that the role of the specimen is simply to wait, ideally in an adequately preserved state, until there should be some need to consult it. This need could be the requirement to compare a new find with an existing specimen for identification purposes. In most cases, however, the specimens will await the interest of a systematist either reviewing the classification of a group of organisms or reviewing the species found within a particular geographic area. Curation of specimens may happen on a rota basis, but often it is not an active process of looking at specimens and checking their current conditions. Rather, when a cupboard is opened to retrieve a specimen for a particular purpose, other specimens in the cupboard may be quickly reviewed to see if they need care, for example to avert insect damage or top up alcohol in jars.

Collections come into their own in the preparation of the two key products of the systematist, namely the preparation of the taxonomic monograph revising the classification of a group, and the preparation of a flora, a fauna, or (in the case of fungi) a mycota, detailing the species to be found in a defined geographic region. In both cases, but particularly in preparation of a monograph, the systematist undertaking the work will examine specimens of each of the species concerned, often looking at many different specimens of the same species, while always alert to the possibility that previously accepted species or genus boundaries may be questionable. The collection provides the possibility for the systematist to sit at a bench with representatives of many species laid out before her, and to compare one with another. This is facilitated in many instances by the way in which specimens are stored: herbarium sheets can be collected and laid out on view, insects are often stored in drawers with many individuals from a species laid out in rows. On this basic activity of reviewing and comparing specimens rests the foundation of systematics as a material culture. Collections are there to bring the variation of living organisms within the view of the individual systematist. As Wyse Jackson (2003) described it in the quotation with which this section began, the collection becomes a library of life for the systematist, from which specimens can be selected and compared. A similar style of description is employed by the Royal Botanic Gardens at Kew: "In the Herbarium, specimens are arranged systematically in the cupboards by family, region, genus and species, so that anyone can find an example of a particular species within minutes. It is, as it were, a card index box of the world's plants, all 400,000 plus species, in a single building but with the sheets arranged systematically to reflect affinities and, usually, evolutionary relationships, rather than alphabetically" (Royal Botanic Gardens Kew n.d.-a). The value of the comparison lies in the acceptance of specimens as proxies for living organisms in their natural habitats. This ability of the herbarium collection to represent the naturally occurring distribution and variability of plants enables use of the collection (rhetorically at least) to substitute for travel: "In the herbarium one can travel from Maine to Alaska and California and back again in five minutes" (Cronquist 1968: 18). An herbarium also allows the botanist to transcend time, both in the sense that the historical variation in plant forms can be studied, and in the use of herbaria to lodge voucher specimens that allow verification of the plants involved in a particular published study at a later date. The system of voucher specimens gives a foundation to tie nomenclature back to particular material artifacts. Specimens in the collections of systematics remain relevant potentially forever: there is no point, short of complete destruction, at which it can be said that a specimen is of no further use.

The distinctiveness of this eternal relevance of specimens in systematics can be illustrated by a contrasting scenario taken from genetics research. In a genetics laboratory where I was based as an ethnographer on a previous project, I spent some hours shadowing a doctoral student engaged in a lengthy and complex procedure. The separation of DNA fragments by size on an electrophoresis gel requires patience, dexterity, and accuracy. It was a great relief to me when, at the end of the day, we viewed the gel in a dark room and found that the procedure had worked: our gel had separated out fragments that could clearly be distinguished by size. I was shocked, then, when the gel was simply crumpled and swept aside once it had been photographed. I had expected that something which had taken such effort to produce would be more precious. In that instance, however, the material object was of minimal importance. The genetics laboratory dealt in information, and the electrophoresis gel was an expendable step in a process of information generation.

This is not to say that the genetics laboratory had no precious artifacts. Stocks of DNA from carefully designed crosses between mouse species formed the basis around which an entire program of research was planned. However, the relevance of these stocks was limited. Beyond a research program of a few years at most, these stocks of DNA would be used up and research interests would move on. Genetics may also develop a longer-term material culture. Recent initiatives in "biobanks" have focused on long-term cryogenic storage for biological samples from humans accompanied by the data on health and environment that would allow them to be used for a range of as yet unspecified research projects (Watson 2004). In this regard, then, the material culture of genetics appears to be moving toward that of systematics. As yet, however, the biobanks are not as embedded in the practice of genetics as their counterparts in systematics. As shown in this chapter, systematics has well-established conventions for the ownership, loan, storage, preservation, and access to its specimens: all conventions that are in the process of being worked out for biobanks. And indeed, whatever the systems that emerge in genetics, it is clear that they will be different in that they do not have to encompass a legacy of historical specimens to the extent that systematics does.

Bowker (2000) makes a related point about the relationship between scientific knowledge production and data. The convention lately has often been that journal articles are the scientific record, and the data that support the theoretical position developed in the article are relatively disposable. Long-term openly accessible archives of scientific data are a relatively new phenomenon, involving a shift toward seeing data as a product in its own right. Bowker considers this a new phenomenon, made apparent in such developments as the Human Genome Initiative. It is certainly fair to say that the database as a part of the scientific publishing system is a relatively new arrival. We might alternatively, however, think of taxonomy as a science that has long been archival. Holding on to the materials on which past taxonomies were based is routine good practice in this field, and in this context the shift to databases as ways of storing data about those specimens is not so radical. The archival culture of taxonomy, in this regard at least, transfers relatively smoothly from the material to the virtual.

Thus far, specimens have been discussed as if all are equal. Within systematics collections this is not true in one very important respect. Certain specimens have an importance beyond others as the "type specimens" on which the description of a new species is based. Daston (2004) gives an in-depth introduction to the ramifications of the type system and the negotiations that went into its establishment in nineteenth-century botany. Here I will only briefly introduce the workings of the system. On the occasion of description of a new species a type is designated. This is not intended as a "typical" member of the group. Rather, it is the specimen to which any decisions about revision of the grouping or the application of names will refer back, the voucher specimen for a species description. Having type specimens is intended to reduce ambiguity by making it clear what the authors of taxonomic revisions had in mind. The importance of type specimens in collections is marked out in physical ways. They may be kept separate from other specimens, and their folder or jar may be marked with a distinctive color.

The classic kind of type specimen described above, where the author of the new grouping designates the type, is also known as the holotype. These holotypes inevitably become lost and damaged, or there may be ambiguity over which specimen was originally intended owing to poor documentation or the original author's having neglected to designate a type. A long-running taxonomic joke asks which person should be considered the type specimen of *Homo sapiens*. Though Linnaeus described the species, he did not designate a type: should he then be considered the type, given that he obviously studied himself most closely, and if so, what should be done about the lack of specimen? In fact, the nomenclatural codes that legislate the process of naming taxa lay down a set of procedures for designation of types in the absence of a holotype, although the application of these procedures, as with many aspects of the nomenclatural codes, can lead to extended discussions. A set of alternative kinds of type has gradually evolved to describe the various situations under which new types may be designated and their possible relationship to the holotype. These include the lectotype, designated by another author, but chosen from materials that the original author examined, and more flippantly, the kleptotype, a piece apparently stolen from the official type specimen and kept in another collection.

Under current botanical rules, an illustration is allowed to be designated as a type, but this is permitted if and only if it is impossible for an actual specimen or part of one to be preserved, which is a condition rarely satisfied. A permissible example might be a case where the holotype has been accidentally destroyed but a clear illustration of it still exists, and might be referred to if there were doubt over the application of the name. A concern to acknowledge historical precedent would thus take precedence over doubts about the adequacy of illustrations. Current rules acknowledge the possibility of visual representation standing in for an actual specimen, but the general superiority of the specimen over illustrations is still asserted. The type specimen, as the material artifact to which the taxon name is attached, is thus reinforced as highly important in the practices of taxonomy. It is likely that anyone who wishes to review or revise the classification of a group will need to examine the type specimen. There are well-established systems by which taxonomists revising a group may be lent specimens from collections around the world. Type specimens may be lent less readily than other specimens, or indeed there may be restrictions that exclude the participation of type specimens in loan systems altogether. In this case, the only alternative is to visit the collection to examine the type specimen.

Specimen collections are considered the property of the institution in which they are lodged (Owens and Prior 2000). There is a well-established system of specimen loans, by which a systematists seeking to study or revise the classification of a particular group of organisms would apply to borrow specimens from other institutions. The Natural History Museum in London (formerly the British Museum [Natural History]) has participated in lending out specimens only since 1902, when formal government permission was sought to overturn the previous strict understanding that collections were for use within the museum (Thackray and Press 2001). There was, however, already an accepted system for loan of specimens in Europe. Such is the scale of this practice that tracking of loans currently provides one of the major rationales for collection databases, incorporating bar-coded specimens that are swiped on sending and on return. The specimen collection looks, from this perspective, even more like the conventional library. By borrowing the appropriate specimens from other collections, each taxonomist is able to assemble a set of specimens as a personal scientific instrument for viewing variation in the set of organisms that interest them. The geographically dispersed network of collections forms a kind of "meta-instrument" from which temporary, localized instruments are assembled, and putting collections databases online makes sense as a way to augment this existing loan system. Using online specimen collection databases makes sense to systematists who are used to tracking down specimens in far-flung locations to constitute their working instruments.

Specimen Images and Working Practices

Thus far, the main aspect of virtual systematics which has been discussed has been the collection database, an electronic version of the catalog which would allow tracking of the specimens in the collection, and which might give improved access both to specimens and to the information about them. There are, however, some uses of information technology that claim to go beyond the straightforward catalog to explore in more depth the potential of virtual specimens to replace altogether some of the functions of the material specimen. A World Wide Web search for "virtual herbarium" returns numerous examples: using Google (http://www.google.com/) on February 17, 2004, returned 5,510 instances, including no doubt many duplicates, but giving some sense of the popularity of the term. The results combine Web sites set up by private individuals and the Web sites of established systematics institutions. On the occasion when I made this search, the top result was a series of photographs of the plants of the Appalachians, produced by an amateur enthusiast (Stein 1994). The second result was the Botanical Resource Centre of the Fairchild Tropical Garden including their "virtual herbarium," constituting "a huge advancement in herbarium use and design coupling the collection of physical specimens directly with the WWW and incorporating complete specimen data integrated with multiple resources for information generation and retrieval" (Fairchild Tropical Garden 1999). While this example appears largely directed toward a professional audience, others, such as the Plant Information Centre of the University of North Carolina Herbarium (n.d.), have an explicitly educational role aiming to reach out to public and schools.

In the instances described above, the term "virtual herbarium" implies the inclusion of images. In the fourth result on the search engine list, a further connotation of the term is found. The Australian Virtual Herbarium (Council of Heads of Australian Herbaria 2003) is an initiative to provide seamless access to the holdings of all the publicly funded herbaria in Australia. A similar initiative in Europe, ENHSIN (Scoble 2003), aims to provide access to specimen databases across Europe. A virtual herbarium of the Chicago region combines specimen data with images of specimens and labels (Schaub and Dunn 2002). The emphasis in the Australian Virtual Herbarium is not so much on the inclusion of images, as on the access through a single portal to information on the holdings of several herbaria. A feature well beyond the facilities of a simple specimen catalog is the inclusion of map information, allowing the user to view placement of herbarium specimens for a chosen species on a map of Australia, coding each by the herbarium that holds the specimen. Though not focusing on virtual specimens as such, this feature certainly reinforces a view of the herbarium as a means to explore natural variability of organisms in space. This form of virtual herbarium ties together geographically distributed collections in a way that emphasizes their collective role as an instrument for exploring biodiversity.

Exploring the list of virtual herbaria reveals various combinations of the different formats. The mixture of amateur Web sites and institutional Web sites includes varying balances between specimen databases, access to further information such as distribution, images of specimens, and photographs of organisms in their natural habitat. In the main, however, the virtual herbaria that will be of interest to systematists, aimed explicitly or implicitly at professionals and serious amateurs rather than the lay public, are the sites of the major botanical institutions. Here we find a variety of strategies exploring the possibility of online information and digital images to supplement, and in some instances replace, the functions of the existing herbarium. In chapter 3, I discussed the theme of the adequacy of digital specimens. In these virtual herbaria, we find such issues being worked through in practice. A co-construction of virtual and material is in process, as the development of virtual resources entails the recognition of new qualities to be valued in material specimens and the information now recognized as being locked away in their labels.

The virtual specimen does have some obvious appeal in the context of the drawbacks of physical specimens. A virtual specimen is infinitely copyable and shareable, does not deteriorate with use, does not take up appreciable physical space, and should be long-lived. The Keeper of the

Herbarium at the Royal Botanic Gardens, Kew, maps out the potential future of herbaria as involving use of virtual specimens to substitute for the real:

> The imaging of herbarium specimens could lead to a revolution beyond that of greater and faster access to the image and its associated data. Herbaria of the future could include specimens held in underground bunkers filled with inert gas (nitrogen, carbon dioxide), and kept at a low temperature with humidity control to reduce the risks of fire and the ravages of destructive insects. Most scientific work would be carried out on electronically held data and images, and rarely might the actual specimen be required. These bunkers could be monitored constantly and if specimens needed to be seen they could be collected, delivered and returned by robots. (Owens 2003: 27)

Type specimens were identified early on as a priority group for cataloging by database, to facilitate location of required items across herbaria (Shetler et al. 1971). More recently, in a context where type specimens are embedded at the heart of taxonomic practice but may be restricted in circulation or use because of their precious nature, digital images have found a niche. The need to examine type specimens forms a viable public rationale for digitization, combined with institutional concerns with displaying their wealth of type material while protecting it from unnecessary loan or casual use. When interviewed, however, many of those concerned with the digitization of specimens, including type specimens, remain unclear about how they will actually be used by systematists.

Possible uses of images include examination in order to verify whether a loan request or visit is justified, but there is little confidence as yet that images can be used as primary sources for taxonomic work. Moreover, the enduring value of physical specimen collections has also been advanced in direct contradiction of this point: "Regardless of how much information in museums is databased or how many specimens are scanned and high-resolution images posted to the World Wide Web, the ultimate value of collections resides in specimens" (Wheeler 2004: 578). Wheeler makes this point on grounds not easily reducible to practical qualities of either kind of specimen: "There is something deeply significant about being in the presence of the actual object" (ibid.). In more pragmatic discussions, justifications for the provision of images often speak of identification, rather than use in the preparation of taxonomic revisions. I invited comments from members of the Taxacom mailing list on the uses they made of online specimen images. Many responses were enthusiastic, but descriptions of the function of images mainly revolved around identification or selection of specimens to request for loan, rather than any direct replacement of the material specimen.

There are, then, some cautious notes to add to the brave visions of indestructible virtual specimens as the proxies for their more fragile material counterparts. There is a deep uncertainty about whether and how the virtual specimens, and indeed the catalogs of specimens that are being made available online, will be used in practice by systematists, if, indeed, they will be used at all with any regularity. In many fields, such a lack of clear analysis of user requirements and rationale for systems design would be a marker of disaster. This is not so in the current climate of systematics, for two key reasons. First, the political environment demands that this information be made available, and it would be extremely difficult for systematics institutions to resist that pressure. Second, the material culture of systematics is an archival one, in which objects are kept for their potential use rather than any known instances of actual future use. In that context, the concerns about systematic databases revolve around making as faithful as possible the representation of the underlying content of the collection, rather than in defining in any detail the ways in which the database will be used. Fidelity to the content of the collection takes on two aspects: first, the representation of as many of the important features of the specimen and its labels as possible; and second, the avoidance of undue prominence for features that might be misleading, such as ambiguous annotation or unpublished names. In this context, then, detailed analysis of requirements from the perspective of potential users is largely irrelevant.

A database is a representation of the collection, and the concern with creation of such databases is to provide an adequate representation of the qualities of the collection as they are valued in systematic practice, although these qualities will not necessarily have been articulated in advance but will emerge once the topic of databasing arises. A collection of specimens is itself a form of representation, and is often portrayed as such by systematists themselves. Already associated with the collection as a form of representation is a whole array of different forms of re-representation of the collection, such as monographs and synthetic works, card indexes, identification keys, and illustrations, each with its own set of conventions and routine working practices. Adequacy of these forms of representation exists through their insertion into particular sets of practices and as expressed for particular audiences. New representations can occasion the specification of the valued properties of old forms.

Examining the status of illustrations as a form of representation in systematics provides a useful basis for positioning the potential of digital images in systematic practice and for reflecting that the adequacy of representations is an issue on which the discipline has some well-rehearsed positions. As an undergraduate botanist I was taught that it was the mark of an amateur to use an identification manual with pictures. True and reliable diagnosis could only be carried out using a dichotomous key, which systematically led one through characteristic by characteristic to an identification. Pictures could seduce one away from this reliable process through mere resemblance, leading one to make a misidentification of a specimen with one that looked superficially similar in the picture. Daston (2004) describes Linnaeus' injunctions that adequate botanical illustrations should mirror plants, but only insofar as they portrayed the important features for classificatory purposes, and particularly should exclude color. The ambivalent attitude toward pictorial representations comes through in the classic text on the flora of the British Isles by Clapham, Tutin, and Warburg: "A volume of illustrations is in course of preparation but as it cannot be ready for some time yet, reference to illustrations in easily accessible floras have been included wherever these drawings were sufficiently satisfactory to be a real aid to identification" (Clapham, Tutin, and Warburg 1962: xiv). The tone seems to suggest that the authors often found themselves disappointed in drawings that were lacking in the qualities needed for identification.

Botanical illustrators are expected to portray the taxonomically relevant features of a specimen, without drawing undue attention to any peculiarities of that individual specimen that might confuse. In the words of the Kew Herbarium Handbook: "Remember a 'stiff' but accurate drawing is scientifically preferable to a 'lively' but inaccurate one" (Bridson and Forman 1998: 152). To this extent, a skilled botanical drawing has often been held superior for identification purposes to a photograph, which cannot help but include the particularities of the individual specimens rather than portraying the ideal version of the taxon. In this, current systematics to some extent preserves the link between idealized images and objectivity that Daston and Galison (1992) describe from the late nineteenth and early twentieth centuries and does not subscribe to a preference for mechanical reproduction. The role of photography continues to be a matter of debate in systematics, just as in ornithology as described by Law and Lynch (1988).

In a context where even photography is controversial, it is no surprise that digitized images were not immediately embraced for all purposes. The potential uses of images were treated with some skepticism on the Taxacom list. Indeed, an early attempt to use the list to invite long-distance identification of a specimen via its digital image was treated with some hilarity. In December 1993 the invitation to identify a fish specimen via an image on a gopher server (reproduced as figure 4.1) produced a long thread of jokes and comments as well as discussion about the technicalities of producing and

==

Date: Wed, 8 Dec 1993 18:49:41 -0500
Reply-To: Julian Humphries <jmh3@CORNELL.EDU>
Sender: Biological Systematics Discussion List
<TAXACOM@HARVARDA.BITNET>
From: Julian Humphries <jmh3@CORNELL.EDU>
Subject: Can you identify this African characin?

As something of a test I am interested in the power of the Net to assist me
in everyday curatorial tasks. So this is somewhat of a test of the
mechanism, although the actual problem is real.

I am curious to see both whether there are people reading this who can
answer such questions as species identifications and whether the method of
depositing a image on a gopher server is a good way to supply the "raw" material.

I have a characin from northeastern Zaire, state of Equateur. I have tried
keys to Western Africa characins with little success. Gery is almost
worthless for this kind of thing. I am hoping that this species is common
enough that someone may recognize it directly from the picture which is
located on the biodiversity gopher at huh.harvard.edu (port 70). The file
is called characin.gif and is located in collection_catlogs/by_institution/cornell.

The fish has biserial premax. teeth, but I am not sure what else is
important. Let me know if more characters are needed.

Oh, the picture is only fair because I simply placed the specimen directly
on my HP scanner and grabbed a grey scale image. I think if I played
around a bit with the specimen and image inhancement I could get a slightly
better picture, but too much time and this kind of thing becomes self-defeating.

Let me know what you think or send an identification if you can.

Thanks,

Julian Humphries
The Vertebrate Collections and The MUSE Project, Cornell University
Building 3, Research Park
83 Brown Road
Ithaca, NY 14850
Voice: 607-257-8143
Fax: 607-257-8109
Email: jmh3@cornell.edu

==

Figure 4.1
An early use of the Internet for taxonomic identification.

displaying images and concerns as to whether one could make a useful determination via a digital image.

The request was framed explicitly as part of an experiment to see how the Internet could help with the everyday tasks of being a curator. The production of such images seemed to be remarkable, given the number of responses inquiring how the fish had been placed on a flatbed scanner, how it had been cleaned, and whether the scales might clog the ventilation slots! Some responses also expressed some disapproval of the use of the list for these purposes, implying that the author should have made more effort to identify the specimen locally or used traditional routes. These discussions mark, as much as anything, the novelty of the use of email distributions lists at the time. They also make clear that although digital images held an attraction for many at the start, there was work to be done in these initial stages to achieve both the acceptance of the digital image as a viable site of identification and the establishment of the working practices to allow one to invite the assistance of remote colleagues in this way.

In time, the use of the Taxacom list to ask for help with identification by posting a link to a Web site carrying an image has become routine. Certainly, such a request now is not met by the jocularity and number of responses that the initial request in 1993 received. This is some indication that images have become accepted as sites of identification, although it remains open whether the attributions that list members make are ones they would stake their professional reputations on. Such identifications as are offered are often not unequivocal, but framed as suggestions or statements about what a specimen probably is not. An image is seen as a site for saying something sensible about the identity of the specimen, even if not for making a solid determination. The use of images for identification is not necessarily, however, an indication that they can be used for serious systematic work. A virtual herbarium with images may allow a systematist to explore some qualities of specimens online, and in particular to research possible visits or loan requests, but the virtual specimen shows no signs of replacing the qualities of the real specimen in many aspects of systematic practice. The virtual herbarium may help to tie together the taxonomic system as a scientific instrument, but it does not appear as yet to be substituting for the material parts of the system to any large degree.

There are exceptions to this picture which depend on particular qualities of the organisms being studied, the means of preservation being used, and the characters that are useful in taxonomic work. In response to my questions about use of specimen images, a fish biologist on the Taxacom list described the very particular qualities of fish images as compared with fish specimens:

In many fishes—and in coral-reef fishes in particular—the quickest way to know what species it is, or at least narrow it down to one of several species—is to look at the life color. Fish specimens are preserved in alcohol in jars, and unfortunately, the life color is almost entirely lost after a few weeks in preservative. Our photos are taken when the specimens are fresh, before they are placed in preservative, and thus the images are the only record of the life color of the specimens. Thus, often times comparing with the images is the best/fastest/most reliable way of confirming the identifying other specimens.

He also described the way that using the fish image had become a more convenient working practice that gave him almost everything he needed in an accessible form:

And it takes a lot less time to call up the images on my computer screen, than it does to get up from my desk, walk into the collection, find the actual specimen, pull it off the shelf (sometimes requiring a ladder), and take it out of the jar (not to mention the need to return it properly to the jar and the shelf, and the potential damage to the specimen itself), etc., etc. So, even though the specimen itself is only a few dozen feet from my desk, it's still MUCH easier to look at the image; and in many cases (probably close to 80–90%), the image gives me the information I need.

However, he noted the difficulty of making reliable measurements from the computer screen. This example demonstrates vividly the co-construction of digital specimen images as the right tools with particular conceptions of the job in hand, the working practices to achieve it, and the accessibility both of the specimen and tools to work with it.

The qualities of digital images have been negotiated in the face of existing practices for imaging. The Royal Botanic Gardens at Kew have a history of producing Cibachrome photographs for imaging specimens, often as a way of sharing information on type specimens held at Kew with herbaria in the plant's country of origin. This practice has, in some cases, been more popular than a loan of the specimen itself, since the Cibachrome can be kept by the receiving herbarium and stored for future use. In arriving at a procedure for producing digital images the concern was to produce a comparable quality and usability to the Cibachrome version. Again, we see that imaging technologies find their value among particular sets of practices, rather than holding those qualities in themselves. Whereas a photograph might be controversial in an identification book, for an herbarium the choice between a temporary loan, a digital image to download slowly through a temperamental connection, or a high-quality photographic image on hand forming part of a cumulative collection could be an easy one to make.

The value of digital images is therefore contextually variable. The technological solutions for producing specimen images are also diverse and highly specific to the job at hand. The production of images is laborious, and each type of specimen requires a different solution. The particular qualities of herbarium specimens have prompted some custom-designed approaches to imaging. The solution adopted by the Royal Botanic Gardens at Kew involves turning the scanner upside down in a cradle over the specimen, in order to minimize the risk of damaging fragile specimens (Royal Botanic Gardens Kew n.d.-b). A Darwin Centre Live presentation on "Virtual Collections" (Pitkin 2002) described a range of the solutions, including conventional photography, scanning of specimens, use of video, and scanning electronic microscope, discussing which was appropriate given the size and qualities of the organism concerned. The problems of field depth when photographing small beetles through an optical microscope were described, together with the software that could overcome these problems by patching together multiple images. The production of images was presented as a skilled craft, involving appropriate solutions developed over a period of time and experience. The Darwin Centre Live presentation on virtual collections concluded with a demonstration of an image of a zoological specimen in a jar that could be rotated on screen with the cursor and viewed from all angles. A member of the audience asked how feasible it would be to digitize all of the museum's collections in this way. The presenter cited an estimate that it would take 350 years simply to catalog all of the collection by database, without any imaging!

In the Darwin Centre presentation, some of the collections available currently on the Natural History Museum Web site were demonstrated and potential uses were invoked: to explore a specimen as a potential loan request; to make use of an interactive key; to identify by comparison; to explore variation within a species. The perceived advantage of the 3-D navigable image was in its reproduction of the activities of turning round a specimen for examination, offering users a sense of control and not restricting them to the angles chosen by the photographer. Solutions for viewing images vary. Some collection Web sites simply display a single image at a time, while others provide facilities for moving around an image or zooming in on portions of interest. In this regard, systems for the display of images have attempted to capture features that will be of interest to systematists and to mirror as far as possible the ways in which they would work with specimens, while also trying to make the procedure as efficient as possible and avoid damage to specimens. Ultimately, however, the adequacy of imaging technologies is negotiated in relation to particular tasks and working scenarios, and no single solution is likely to please all.

The Organization of Digitizing

It is clear that digitizing specimens, whether in producing a catalog database or providing high-quality images for taxonomic work, requires considerable labor. Chapter 2 demonstrated some of the ways in which digitization projects have become essential activities for systematics institutions to engage in if they are to portray themselves as forward-looking and as meeting their obligations to share resources. This has not, however, always been the case, and in the past the decision to digitize has sometimes been controversial. Collections of specimens are often described as stores of information in their own right. In an introductory text on plant taxonomy, Jeffrey (1982: 31) explains the function of the herbarium in this way: "The herbarium is, therefore, a data store in which information about plants is stored mainly in the form of dried plant specimens and their accompanying field notes." In similar fashion Massey describes the herbarium as "a data bank with vast quantities of raw data" (Massey 1974b). So impressive is the herbarium's ability to act as an information source that serious doubts were expressed in the late 1960s on whether computerization could add substantial value: "The suggestion is sometimes made that it might be desirable to computerize the whole herbarium. It seems easy to overestimate the potentiality of the computer and I am far from convinced that a good well-managed herbarium does not incorporate in its own arrangement its own index, as easy and as efficient to use as any provided by mechanical means" (Cronquist 1968: 20). Others at the time considered computerized records at least worth experimenting with: McNeill (1968) cites the work of the Museum of Natural History at the Smithsonian Institution approvingly in this regard.

The sheer amount of specimens was represented as both the problem and the motivation where computerization was concerned. Many authors pointed out that it was simply impractical to expect herbaria or natural history museums to be able to enter data on their entire collections into computer. Precisely because of the bulk of material, however, computerization was seen as a solution to concerns about accessibility. In addition, developments in taxonomy meant that observation of a preserved specimen, while still key, was only a part of the process. Chemical and genetic information began to play a part and posed new information retrieval problems, as Heywood (1976: 50) describes: "The present apparatus of taxonomy, comprising the binomial system, Latin nomenclature and the hierarchical arrangement of categories, provides one of the best information retrieval systems yet devised by man. Yet the sheer bulk of the rapidly accumulating additional information of all sorts which is relevant to taxonomy is such that electronic data-processing equipment will be required to handle it so that it can be made readily available for classification and other systematic applications." As the political climate has shifted, and as new technologies have emerged for imagining the role of the specimen collection and the location of its potential users, the rationale for digitizing has become cemented as the self-evident truth we saw in chapter 3. How this digitization is to be organized and how to address some of the problematic issues it brings to the fore have become issues of daily concern and public debate, which disaggregate the blanket rationale for digitization into concerns about particular audiences and sets of priorities.

The databasing of specimen collections has involved some uncomfortable choices, particularly on whether everything available from the specimen and its labels should be entered into a database. The current trend to make databases available on the Internet has exacerbated these concerns, focusing in particular on the inability to control the audience for information available on the Web. While it is generally felt desirable to make information accessible, some concerns focus on improper use of information on the locality of rare plants. Such information may be deliberately excluded by an institution from its publicly available databases. Other concerns focus more directly on the systematics community. Fears here revolve around circumstances where label information may not equate to validly published taxonomic information. A working systematist might have labeled specimens with new names that were never, in the end, published in the formal literature. Some discussions on the Taxacom mailing list revolved around the concern that making these non-names available in publicly distributed databases could promote the use of names that should not be in circulation. Similarly, there was some debate on whether to include in databases annotations noted on a specimen by a systematist working with the group, where there may be little clue as to what the author meant by the annotation. A post to the Taxacom list parodied the resulting confusion:

> Through the magic of net technology, we bring you the thought processes of one 21st Century taxonomist as she scans a database:
>
> Hmm, what's this? "_Snailus_ sp., aff. _S. albus_." Shit. What's that supposed to mean? Let's see who ID'd it. Barry Roth. Shit. He was so enamored of that silly principle of grouping by synapomorphy (John McNeill's later papers put an end to that foolishness). So I guess when he wrote "aff.," he meant that this specimen shared some derived character with _S. albus_ but had some unique features as well. Let's see the scanned image of the label—good thing this database stores binary objects (good ol' Borland Co.). Looks like Roth originally wrote it in pencil. Shit. He must not have been too cer-

tain. Why didn't he just use a "?" Oh wait, that was around the time of the XLVIIIth Congress (the "Jensen" Congress) that tried to outlaw question marks in taxonomy. Well-meaning, but human nature is human nature. Besides, how can you codify uncertainty? Shit. Too bad I can't see the specimen. The Academy of Biological Sciences sold off its collections to make room for interactive displays. That was right after NSF went bankrupt (they should have listened to Jim Croft). I don't suppose Pepsico even kept the records of sale. Shit. (Roth 1993)

Paradoxically, then, the fears expressed with regard to making specimen databases available online revolve around the concern that they may make things "too real." By being thought erroneously to replace real (physical) collections, if taken without sensible reference to their context of production, and by making accessible some features of real collections that ought to remain hidden, the fear is that databases could disrupt the existing working practices of systematics. The situation is summed up by Guala (2000) as "All ghosts come out of the closet": a suggestion that the creation of online specimen databases may cast an uncomfortable light on the classification, curation, and labeling practices of the past. The accessibility the database is seen to provide then becomes double-edged: the process of databasing can force decisions on matters previously left quietly unresolved, and require, for maximum effectiveness within the database, that ambiguous matters be made artificially clear.

This issue makes clear that digitization requires skilled labor if it is to attend to the nuanced issues important for making the right kinds of information accessible in the right ways. Maintaining databases has become an accepted part of the curator's labor in looking after collections of specimens. Advertisements for curatorial positions posted to the Taxacom list in 1993, the first full year for which archives are available, ask for skills that include: "on-the-job experience with computer data-entry or ability to learn to use a computer"; "familiarity with botanical specimen databases (both concepts and practices)"; "solid understanding of organization and implementation of computerized database systems and their applications to museum collections"; "experience with the microcomputer applications (including database management and design, electronic networking, and routine systems management for DOS & Macintosh platforms)"; and state "computer skills and experience with field collecting methods are important." Not all curatorial positions at this stage specify the need for skills in computing and specimen databases, although in some of those that do not, it is possible that databasing is implicitly included in their specifications of the need for curatorial experience. Advertisements for positions posted to the list in 2002 show that explicit mention of the need for experience and skills in databasing and software use is still commonplace. Where it is not stated explicitly, it seems that activities such as processing incoming and outgoing specimens, documenting collections, and keeping records are now assumed to include the use of databases: an advertisement might, for example, state under "job responsibilities" that the applicant is expected to maintain collection documentation, and only under "qualifications required" make explicit that experience with database packages and familiarity with computers is expected. From these advertisements, even though they do not constitute a sufficiently large sample for any sensible conclusions about trends, it is possible to see that the maintenance of computerized records of collections has become a routine part of existence for many collections managers and curators. While some institutions are in a position to advertise for such specialist staff as "Digitization Manager" or "Bioinformatics Distributed Information Systems Programmer," in the majority of cases the tasks of digitizing collections are part of the routine responsibilities of curation.

As we saw in the previous section, the production of images is also laborious, adding yet more to the obstacles in the way of realizing the complete accessibility of collections. In addition to the specimens and their labels, many collections already have card indexes, which contain a wealth of taxonomic information that could have value for distributed users. Again, just as for specimen label data, the labor of digitization is a major obstacle. One innovative solution, developed in the LepIndex project at the Natural History Museum in London, uses a modified bank check scanner to produce images of cards, then combines image analysis, optical character recognition, and use of dictionaries in order to automate as far as possible the process of database production (Beccaloni et al. 2003). A key point to note is that the resulting database retains the images, allowing a picture of the original card to be consulted. Users can add their own validation of the data to the archive. This system thus incorporates the expectation that the products of digitization will be imperfect and enlists users into the project of ongoing improvement of the data.

Decisions about what to digitize are a significant site where specific kinds of labor may be demanded in order to make sure that accessibility, often presented as a universal benefit, remains a positive gain. Far from replacing other features of herbarium work, programs of digitization have largely been additional to existing work even while incorporated into the job descriptions of collection curators. Within this context, it is not surprising then that digitization has also been an occasion for prioritization. The priorities for making taxonomic information available in digital form focus on the differentiation of specimens, the differentiation of taxa, and the availability of funding, three not necessarily independent dimensions. The differentiation of specimens involves selecting particular kinds of specimen as most important to capture in digital form or make available online. In this category fall, most obviously, type specimens. The differentiation of taxa depends on identifying groups of organisms that could most usefully be made available, often those that are economically important or crucial indicators for conservation purposes. The final distinction focuses on the fundability of projects: the availability of Darwin Initiative funding to make UK based taxonomic expertise available in less developed countries has often been used to fund projects that, at some point, involve making specimen data and/or identification resources available online. This, in turn, often also depends on the characterization of the relevant taxonomic group as a priority from a conservation perspective. It might also depend on the perceived importance of a particular set of specimens, where funding is available to make the type specimens of a priority group available to taxonomists in their nation of origin.

This formulation of priorities is not an inherent part of the digitization project. One could, quite conceivably, simply begin at the beginning of an herbarium or natural history collection and work through each specimen in turn, scanning, photographing, and inputting label data. This would, however, be a long-term and labor-intensive process that would lack the important qualities in the current funding climate of a demonstrable product in the short term. The prospect of complete digitization is important, as we saw in chapter 3, as a way of organizing expectations about taxonomy's role within biological science and its future position in the global politics of biodiversity. Within the realities of current funding regimes, however, more restricted prioritized projects usually take precedence. Identifying specific user needs for online information services and conducting requirements analyses are not, however, necessarily a part of these priority-setting procedures.

A further aspect of prioritization is that the means of databasing specimens has to be practical from a technical point of view. Not only does it have to provide what is considered an adequate representation of specimen data, and to be feasible in terms of the available data, it has also to be implemented in software as a working system. Over time a wide range of different solutions have built up (Berendsohn 2003a). Figure 4.2, a message to the Taxacom list in 2002 from a participant troubled over making an appropriate choice, illustrates the situation well.

The list of solutions offered to this inquirer includes, along with the solutions developed in-house by individuals from generic database packages,

==

Date: Wed, 16 Jan 2002 20:28:51 +0100
Sender: Taxacom Discussion List <TAXACOM@USOBI.ORG>
From: Roland Eberwein <roland.eberwein@LANDESMUSEUM-KTN.AT>
Subject: Database - summary

Dear Listmembers,

several weeks ago, I posted a call for help: I'm looking for a suitable database to manage a herbarium and a botanical garden.
I got a lot of tips, support and informative discussions - CORDIAL THANKS TO ALL !!
Now I want to share the results:

Many listmembers are working with self-developed databases using MS Access, dBase, Filemaker Pro or Paradox. I don't want to develop an own database because I'm a single fighter in "my" herbarium. If I go mad, it would be difficult to update or convert the database. A "professional" database is more secure: you get support, updates and if the company goes bankrupt, others are affected too and the way out can be easier found by teamwork.
MS Access is the most popular database software (be careful, Access should not be used for larger databases. For example, Willoughby Ass. http://www.willo.com/cgi/content.cgi?main.html,snap.html,intro_left,snap.html,intro_right recommend only 60.000 objects for their Access-based database SNAP!)

Recommended (used by listmembers) software packages are:

[List of eight packages, giving brief notes on source, cost, evaluation of support]

For those, who are looking for a database too, it's worth to open:
http://www.bgbm.org/TDWG/acc/software.htm

Best wishes and
good luck!
Roland

PS: I couldn't make a decision up to now.

==

Figure 4.2
A message to Taxacom, reporting on help with choosing a database package.

commercial packages for management of museum collection data and packages developed within systematics institutions. Many of the latter have been developed as in-house solutions but then generalized and made available more widely, with varying levels of support. The inquirer makes clear in his message that such decisions are affected by a range of issues, including assessments of one's own institutional position and that of the software providers, and the extent to which one is participating in a community of users. Subsequent discussions argued that even attributing technical qualities to software packages, such as number of simultaneous users supported or limits to file size, was far from straightforward. Adding to this the specificity of database packages to particular curatorial practices or arrangements of specimens and specific nomenclatural codes, it can be seen that selecting a solution is a complicated business. Even within an institution diverse solutions may coexist, as individual workers adopt packages they feel most comfortable with, or as different solutions are worked out for the needs of different collections and their curators. A glimpse of this diversity of database solutions for collection management also emphasizes that the role of organizations such as GBIF in aiming to provide interoperability between databases is far from trivial.

The previous sections demonstrated the special status of collections of specimens within taxonomic practice. Within this context, the advent of databases has provided a means of making specimen data and images available outside the institutions that hold them. Crucially, practitioners hold fiercely to a view that digitization does not make the specimens themselves irrelevant. While striving to make digital representations as good as possible, systematists hold to the value of the physical collections as stores of a potential, as yet unrecognized, set of qualities that may be needed in the systematic work of the future. The value of making specimen databases available is expressed within terms understandable to contemporary systematists: researching a potential visit or loan request; returning specimen data to the country of origin; making idenfication facilities available at a distance. In my terms, databases are represented as a means to strengthen the integrity of taxonomy's material culture as a distributed scientific instrument, rather than replacing it. Digitization activities are imagined in terms of the way they fit into and augment existing practices. Detailed justifications of design decisions tend to be made in terms of the adequate reflection of the qualities of specimen collections rather than in terms of meeting particular uses to which they will be put. This is consistent with the view of collections themselves as potential sources of future information as yet unrecognized.

In chapter 2 I described work that stressed the importance of viewing databases as institutionally located. It was vital, these authors argued, to understand how databases were formulated, what the processes were for arriving at data structures and for authorizing contribution of data and access to data. Only by understanding these factors could we arrive at a nuanced understanding of what might otherwise, crudely, be thought of as the "impact" of databases on a social landscape. In the case of the specimen databases described above, the institutional location of the databases largely mirrors the institutional location of the specimens. Just as large institutions are the holders of major collections, so too are they the holders of major databases that describe those collections. The copyright and intellectual property issues raised by specimen databases are complex and have yet to be thoroughly understood by the institutions themselves (Owens and Prior 2000). Experience has shown that databases are thought of as institutional and national property. Although access might be freely given to casual inquirers over the Internet and projects might be undertaken to "repatriate" important sets of data to their countries of origin, data sets are unlikely to be donated wholesale. This renders solutions such as GBIF, focusing on interoperability of databases rather than wholesale merger, particularly appealing since they allow institutional participants to fulfill their responsibility to respond to global concerns while maintaining institutional boundaries and assets. Pressure on finance for the maintenance of collections has been felt acutely for some time, and in this context institutions are alert to the risks and opportunities of collaborating. It is unlikely, in the current climate, that large-scale virtual zoological collections or herbaria will be produced that seamlessly cross institutional or national boundaries without attending to the location of material collections. The pressures on the institutions that hold the collections are simply too great for them to take this risk, and initiatives that do prosper are most likely to be networks of existing institutions, with arrangements to protect the stakes of the participants.

Such possibilities for cross-institutional projects as do arise may be the result of national initiatives, such as the Australian Virtual Herbarium, or may focus on a particular kind of specialist data, like the Digimorph project (http://digimorph.org). This project, providing three-dimensional images of specimens via high-resolution x-ray computed tomography, offers images of specimens from a variety of collections. The production of these images is a specialized skill using personnel and equipment not likely to be available within the museums holding the physical specimens. The data may well be used by systematists, since this specialized form of scanning makes available features of the internal morphology of specimens not available from the external observation of a physical specimen. Such scans have been used in taxonomic publications, particularly in relation to fossil specimens. By providing a specialized form of data, the Digimorph project also develops a new institutional location for specimen data that allows it to sit outside the major institutions holding collections of physical specimens. The new location is organized around the scanning facilities that make the specimens visible, rather than the collections themselves. This is, however, an unusual case.

In the majority of instances the location of physical collections dictates the corresponding virtual collections. Some of the more radical possibilities for databasing in systematics are therefore unrealized and, given the current pressure for funding and political recognition, unrealizable. Paradoxically, by placing systematics institutions under pressure to demonstrate their worth, funding regimes may be stifling possibilities for more radical innovations in the deployment of virtual technologies. This issue is explored in more depth in chapter 6, when I turn to mapping institutions and initiatives that populate systematics as cyberscience. For the rest of this chapter I continue with the qualities of the material and virtual objects that systematics handles, turning next to the ways in which these objects are ordered and looking at the particular practices with material objects that mean that, in the daily work of systematics, the casual equation of digital specimens with accessible specimens does not necessarily hold true.

The Ordering of Objects and Accessibility

The description of the material and virtual culture of systematics above focused on the qualities of specimens and their reproduction in virtual form, and the ways in which digitizing has been organized in response to pressures to make specimens accessible in this form. The key claim being made for virtual systematics which deserves further investigation is the prospect that it is more accessible than the arrangement of objects in space, which is a feature of the specimen collection. In this section I explore in more detail how objects are meaningfully ordered in space in a specimen collection, and how virtual systematics augments and disrupts this arrangement.

One very visible way in which the physical placement of objects is intended to convey meaning is the order beds of a botanic garden. This common feature of a botanic garden distinguishes it from the purely recreational or aesthetic garden. Order beds are often laid out in a geometrical pattern, with plants deemed to belong to the same family situated together in a bed. Some large families may take up a whole bed, while smaller ones will often be combined. Higher-level taxa will be represented by grouping closely related families together. Labels will usually be placed by each species, and at the end of each family group. In this way the order beds become a visual representation of a taxonomic scheme. The species chosen to represent families will be limited, of course, by the climatic conditions prevailing in a particular botanic garden, and some families may have to be omitted altogether. Practicalities also demand that the size and growth habit of species be taken into account in choosing representatives. Nonetheless, the order bed has a long history of use as a pedagogic device, to promote knowledge of plants and to encourage understanding of their relationships.

The order bed is meant to allow viewers to explore the similarities of plants that are placed together, to come to recognize the characteristics of families and orders in a predictive way that is meant to allow recognition of a member of the same family that has not been included in the bed. Brown describes the Muséum Nationale d'Histoire Naturelle in Paris, with its combinations of museum specimens, zoo and botanic garden, thus: "Whether it was stuffed, dried, cultivated or caged, everything in the Muséum was haunted by its own referentiality; because the aim of natural history was to make the living individual point beyond its idiosyncracies to its place in a system of classes within classes like wheels within wheels" (Brown 1992: 62). The order bed, then, seeks to portray the principles, as well as the outcomes, of systematic research.

The arrangement of biological specimens is seen within the context of systematics as a theoretical statement, which has consequences for the way the elements in a collection are viewed and the predictive or pedagogic use an observer may make of them. It is commonplace for a botanic garden to make explicit the source of the classification on which the physical arrangement of specimens is based. The Web site of the Chelsea Physic Garden describes the order beds thus: "The Dicotyledon botanical family beds are laid out in accordance with a modified Bentham and Hooker system of classification. This arrangement was designed to show a possible evolutionary sequence and to indicate relationships between species. This system of classification is based on flower structure, but more recent studies based on molecular analysis would rearrange the beds considerably" (Chelsea Physic Garden 2003). In accounting for the arrangement of the order beds this description stops just short of apologizing for the use of an outdated system of classification. In the early days of the garden, use of an outdated arrangement would have been less forgivable. The garden was set up in 1673 explicitly to help train apprentice apothecaries in plant recognition. Today, while the mission statement still explicitly alludes to an aim with a systematic focus, "to demonstrate through its plantings and publications the range of species named or introduced to cultivation by a succession of distinguished curators," an aesthetic and amenity function is also acknowledged. The teaching of up-to-date systematics is not therefore necessarily as high on the agenda for the garden as it once was. For other botanic gardens, however, a modern classification system in the order beds may be a source of pride and an opportunity to teach visitors not just what the results of taxonomy are, but also how it is done.

The order beds at the Hortus Botanicus in Amsterdam, known as the SemiCircle, are arranged to reflect the classification resulting from application of molecular systematics. The display board accompanying the beds explains their history and the principles of molecular systematics on which the current arrangement is based. Visitors are encouraged to make use of the predictive qualities of the classification to spot the two specimens that are deliberately out of place. The display board ends with the offer of further information:

> For more information on molecular systematics and the new classification of the plant kingdom, please see: http://www.mobot.org/MOBOT/research/APweb/

As I stood in front of the display, on a chilly November afternoon with darkness gradually closing in, I wondered what to do with this encouragement to visit the Web site. I could not imagine that many visitors would do as I did, and carefully note down the URL for future reference.

Later that evening, in an Amsterdam Internet café, I pulled out my notebook, and typed in the URL recommended by the SemiCircle display at the Hortus Botanicus. I had expected the site that filled the screen to be the home of the plant systematics departments at the leading Dutch institutions, Leiden, Utrecht, and Wageningen. Instead, I was taken to a site created by a member of staff at the University of Missouri and Missouri Botanical Gardens (Stevens 2001). The URL at the bottom of the display board transported me from chilly Amsterdam to the tropical warmth (in my mind, at least) of Missouri, from a public display meant to educate "plant lovers" to a site rich in academic references and detailed research information, from a somewhat confusing vista of plants in various states of luxurious or sickly growth, to a text-rich, searchable, and thoroughly ordered set of data. Whereas the order beds showed me a rather confusing array of plants, the Web site gave me a way to see a situation more amenable to my expectations about how information should be ordered. The two were linked by the names they used, but their treatment of them differed dramatically. In the order beds, a label indicated the species and family, with a bar code to link the specimen to the garden's catalog. On the Web site, a name was linked to diagrams of phylogenetic relationships, which linked to formalized descriptions of the characteristics of taxa, and in turn linked to references and occasionally images. Sitting there in the Amsterdam Internet café, I found the site overwhelming. It seemed to contain the answers to everything, but I did not have a suitable question to ask of it.

Reviewing the Web site in more detail at a later date, I found that my first impressions of extreme contrast between the order beds and the Web site were possibly not well founded. At first sight the order bed and the phylogeny Web site seem worlds apart, but they are linked by much more than the display board at the Hortus Botanicus. Stevens (2001) intended his Web site not for systematics researchers, as I had assumed, but for students. The site arose in the context of his efforts to teach students the characteristics of plant families: a task very similar, in fact, to the original order beds of botanic gardens. Far from being copiously referenced, according to its author the site contains only sparse links to literature. He aimed not to be comprehensive, but to provide orderly access to information that students would need in order to understand the evolutionary past of plant families, and to be able to place unknown plants within major groupings. Stevens (2001) also states his intention that the Web site be flexible, allowing for updating of the classification with new hypotheses. The introduction to the site frames this in terms of a comparison with books:

> This series of pages is a set of characterizations of all orders and families of angiosperms (flowering plants), as well as many clades grouping families and orders and some lower-level clades within families. They are designed to help in teaching angiosperm phylogeny at a time when our knowledge of the major clades of angiosperms and the relationships within them are still somewhat in a state of flux. Conventional books are out-of-date before they appear, furthermore, there is no comprehensive phylogeny-based treatment of angiosperms, out-of-date or not. (Stevens 2001)

We might also consider a comparison between the flexible updating of the Web site and the physical labor and horticultural skill required to effect similar changes to the classification of the order beds. We might assume that updating the Web site would be an easy job in comparison, but the description of the process of creating the site suggests otherwise:

> The trees were created by generating a raw dataset in MacClade and drawing the trees in the tree editing window. The tree files were then accessed in PAUP and saved as .pict files to facilitate use in Adobe Illustrator. After polishing the tree images in

> Illustrator, each tree was saved as a .gif file. Each tree image was made into an image map using a tool in HomeSite, which allows you to select any part of the image (treated as coordinates) and assign a link for those particular coordinates to any place within the website or to any other current website. I imagine there are much faster ways and much more sophisticated ways of building a website like this, but I am no expert, so I did it the hard way, by trial and error. As the project got more complicated, I figured out how to do each particular feature. (Ibid.)

The labor of digital systematics substitutes for the gardener's skill. Updating classifications requires a laborious reordering process in both contexts, and is done in both cases for a pedagogic purpose: to represent, through the ordering of objects, their relationships.

The arrangement of living collections according to taxonomic principles appears primarily as a pedagogically oriented choice, but there is also an extent to which this arrangement expresses the garden as a resource from which useful material or information can be retrieved. This aspect becomes most apparent in the description often used of botanic gardens as "living libraries." While the collections will, of course, be cataloged, and increasingly given barcoded labels, databased, mapped and even indexed by GIS systems, the taxonomic arrangement of the order beds gives a quick and accessible overview.

At the Royal Botanic Gardens, Kew, the order beds are located in an area of the gardens off the main path for the majority of visitors, overlooked by the Jodrell Laboratory. In this building research work on plant anatomy, physiology and genetics is carried out, to inform basic taxonomy and more applied questions with commercial, medical, or conservation relevance. The Jodrell Laboratory is located within the gardens. From the perspective of the work done in the laboratory, however, this relationship is inverted. The gardens, or at least the collections they contain, become part of the resources of the laboratory. As material for molecular systematics, chromosome studies, physiological investigation, or anatomical comparison, the plants in the gardens provide a ready resource.

On a behind-the-scenes tour of the laboratory, the group I joined was shown the work of the plant anatomy laboratory where wood fragments could be identified by comparison with a reference collection. In this case the collection was held in two ways, as a collection of slides in a series of filing cabinets, and as a computerized database. Identification might, in the terms used by our guide, mean that "you dial in the characters of the specimen you're looking at and it spits out the identification." Comparison with a specimen from the reference collection would then be used as a confirmation of the identification. While this collection is indexed in filing cabinets according to taxonomic arrangement, the demands of identification mean precisely that one does not know what one is looking for. In order to settle on a specimen for comparison one needs either an expert with a very good memory, or the kind of direct access that a database of characteristics (or any form of identification key) provides.

In this example the reference collection is located within the laboratory, and the identification task means that the taxonomic arrangement of the reference collection is not the most easy to use. The database is an intermediary, allowing access to specimens according to alternative criteria. The Jodrell Laboratory also carries out chromosome studies, investigating the form and number of species. Some of this work is explicitly framed as survey oriented, and here the taxonomic framework within which the living collections are organized gives a useful means of orienting a strategy of sampling. While the laboratory is in the gardens, in these circumstances the gardens are used as a resource within the laboratory. The collections might function as a reference and source of material for research work, but according to the task the exact relationships between laboratory work, collection, and taxonomic classification varies.

Ordering of objects is important to systematics, as a part of the creation of the collection as a scientific instrument. Behind the scenes, in an herbarium or collection of zoological specimens, the conventional arrangement of specimens is by taxonomy. The exceptions to the ordering of collections by alphabet or taxonomy revolve around either the qualities of the specimen, or their involvement in ongoing working practice. A particularly large specimen may be stored out of taxonomic sequence. The behind-the-scenes tour of the Darwin Centre includes a cupboard of specimens grouped together solely on the basis of being very long and thin, and therefore needing a special cupboard to store their tall jars. The "tank room" of the Darwin Centre stores the largest specimens, not in jars but in stainless steel vats of alcohol, the lids lifted by built-in hoists. Memorably, on one of my visits a baby rhinoceros donated by a zoo was being installed in a tank. In the herbarium, separate storage is required for specimens that do not conform to the flat presentation and even size of the herbarium sheet. Spirit-preserved flowers or large seeds would be stored in a dedicated area out of taxonomic sequence with the main collection of specimens. In a well-curated system, there will be a note or dummy folder in the standard taxonomically arranged collection to direct the user to the nonstandard specimens stored elsewhere. The other exceptions to storage by taxonomic ordering are those specimens that are part of ongoing working practice. Specimens on loan to or from other institutions, and part of an ongoing project, may spend considerable time grouped as a collection for work in progress. Similarly, recent acquisitions may spend long periods of time waiting to be identified, processed, and assigned to their correct place in the taxonomic sequence.

The storage of herbarium and zoological museum specimens is therefore organized according to a chosen system, either alphabetic or taxonomic, allowing for exceptions due to the nature of the specimen or the work that requires it. Efforts will be made to maintain the chosen system by leaving space for new specimens to be inserted within the existing order: full cabinets will be the occasion for considerable work in moving specimens along to make space. Whereas the order beds in the botanic garden were arranged in systematic order to make a pedagogic point, the herbarium and the zoological museum collection are where systematics can put its principles into practice for its own working purposes. A taxonomic arrangement can be seen also as a more practical arrangement, as a recent discussion on the Taxacom list illustrates. A query was posted to the list by a member involved in setting up a new herbarium, asking how the collection should be organized. An early response advocated the alphabetical model, but this drew a spirited defense of taxonomic arrangement (figure 4.3). I quote this message in full, for the picture of life as a collections manager that it vividly portrays.

The defense of a taxonomic arrangement made here is that, oddly enough, it is more stable in the face of taxonomic reassignment than the alphabetical system. Given that existing classifications do a reasonable job of reflecting phylogenies, any changes are unlikely to move taxa radically, and thus the taxonomic arrangement is the physically conservative one. The argument was not, it should be made clear, carried by the message quoted in figure 4.3. A number of vehement opinions were posted on either side of the argument: a point interesting in itself for illustrating that even in 2002, with all the possibilities for databasing and searching collections, the physical arrangement of specimens remains an issue of considerable concern to collection managers.

Later contributors to the discussion about herbarium arrangement made links between the processes of arranging a physical collection and of creating a collection database, stressing that the two activities had implications for one another, in terms of the information on organization that the collection database needed to carry and the extent to which the physical arrangement of the collection needed to anticipate particular user needs. It was felt that a database which allowed access to specimens by alternative criteria not anticipated by their physical arrangement would help to overcome the limitations of a single system, but would certainly not make it irrelevant how specimens were arranged. Databases would need to be

Date: Tue, 29 Jan 2002 10:02:56 -0800
Sender: Taxacom Discussion List <TAXACOM@USOBI.ORG>
From: Doug Yanega <dyanega@POP.UCR.EDU>
Subject: Re: How to arrange a new Herbarium?

Under our previous collection manager, we had some 2 million pinned specimens arranged alphabetically by families within orders. Since the original designations 30 years ago, at least 100 family names have been changed - mostly either added or deleted. If we honestly wish to keep pace with all these changes, it would require moving practically every single drawer in the collection. Every time a family name is changed, in fact, for the foreseeable future, we'd have to reshuffle dozens of drawers (since we are almost at capacity, any family that has 5 or more drawers CANNOT be moved without some compensating moves to make space). Consider just one example: the family Scarabaeidae has been split, so in addition to having a section labeled SCARABAEIDAE we'd now need sections labeled BELOHINIDAE, BOLBOCERATIDAE, CERATOCANTHIDAE, GEOTRUPIDAE, GLAPHYRIDAE, GLARESIDAE, HYBOSORIDAE, OCHODAEIDAE, PLEOCOMIDAE, and TROGIDAE (and, if you follow some people, also APHODIIDAE, CETONIIDAE, DYNASTIDAE, MELOLONTHIDAE, and RUTELIDAE). This one taxonomic decision alone would conceivably necessitate the movement of nearly every drawer of beetles in the entire collection in order to get everything alphabetized properly.

This is stupid.

Since we're going to have to move every drawer anyway, we've decided to move them into phylogenetic order by superfamilies, and larger taxa will be divided to subfamily level. This has already been done for the Hymenoptera. This will ABSOLUTELY minimize the number and severity of future rearrangements. In other words, if we'd had our Scarabaeoidea all together to begin with, then all those names above would have already had their own drawers (as subfamilies of Scarabaeidae), and basically all we'd need to do would be change the endings on the drawers from -inae to -idae, and shuffle the Passalids, Lucanids, and Diphyllostomatids in with the others. All the rearrangements and movements of drawers would be confined just to that one superfamily, instead of randomly throughout the whole order. This is a VAST improvement over alphabetizing. Every cabinet of Hymenoptera in the collection has a sheet showing the phylogenetic hierarchy for the order in big friendly letters, with the families housed in that cabinet highlit in fluorescent yellow (and including all archaic and alternate family-level names) so any student who knows the name of a family can easily locate it in the collection... AND they get a lesson in phylogenetic classification at the same time.

Maybe people in herbaria don't mind moving sheets from one cabinet to another every time a family name is changed, but wouldn't it STILL be better if you only had to move sheets at most from one cabinet to an *adjacent* cabinet, instead of one on the other side of the herbarium?

Figure 4.3
A contribution to a Taxacom discussion on the best way to arrange collections.

designed to reflect not only the currently accepted taxonomic system but also the system by which specimens were physically arranged (which, even if it were systematic, would still probably lag behind the currently accepted taxonomic system at least in some respects). A further consideration is that if the physical arrangement were systematic, then an alternative alphabetical listing could easily be produced using the collection database. If the physical arrangement were alphabetical, however, a systematic listing could not be automatically produced by the database. Rather than providing flexible access to the collection, the database in this instance is argued to provide constraints on the sensible physical arrangement (although this point also was disputed).

The promise of collection databases is that they make the collection more accessible. Whereas under the taxonomic system it is impossible to retrieve all taxa from, for example, a particular geographic region, as one would when preparing a flora or checklist for a region, the collection database makes these qualities readily accessible. The collection database provides for alternative orderings of the collection free from the labor of searching cupboards or rearranging specimens. It also, when provided online, gives an insight into the collection holdings for people located elsewhere, who are not able to come and check the shelves for a given specimen. The online collections database thus opens up access to the museum collection in two directions: by providing access to outsiders, and by providing kinds of access that the physical collections themselves cannot. The physical orderings of material objects in systematics are well established, thoroughly theorized and embedded in practice, as the descriptions in this section of order beds and specimen collections have shown. Collections managers have, however, increasingly acknowledged the advantages that databases would bring as alternative means of access to those collections. Collections databases are seen as having a dual role, "to both help in the management of the collections and provide access to collections information" (Natural History Museum 2003: 14). The collection database becomes an object that stands at the threshold of the museum and has meaning both within the museum's practices and to outsiders, becoming inserted into their own sets of local practices.

Networked specimen databases such as the ENHSIN project in Europe (Scoble 2003) or the Australian Virtual Herbarium (Council of Heads of Australian Herbaria 2003) seek to expand upon the threshold function of online catalogs by spinning a web of databases across institutions. Objects remain physically arranged in collections in formats that are crucial to their everyday management and used as insights into the distribution of natural diversity. Unified gateways to collections databases laid across the top substitute for the physical movement of specimens and of personnel, and allow for new activities to be imagined as searches can be structured within collections and across sites by qualities of interest to individual projects. If I want to search out all specimens of a given genus held in major European collections, or find out who holds plants collected in Surrey, or even, in some visions, find all plants that would be favored by a rise in the average temperature, I may in future find a unified gateway to a set of collections database that grants me my wish. The idea of the collections database is to make information previously only accessible from physical specimens available in a form appropriate to these alternative agendas.

Thus far, the advantages of the collections database seem, to contemporary eyes, obvious. There may be some sense of loss for the user, in terms of the awareness of phylogenetic relationships that comes from seeing a collection of physical objects arranged according to their phylogeny, but this is a comparatively minor loss compared to the gains. Collection managers go as far as they can to provide an organized and accessible physical collection, but the qualities of physical artifacts are such that any arrangement will have drawbacks. An intrinsic part of imagining the potential of the virtual is a corresponding expression of the drawbacks of the material. The collection database seems to free up a lot of the labor of reorganizing specimens and sorting through them to find the items of interest, and makes a physically located collection available to a distributed audience.

Distributed collections databases are seen as a way of realizing the full potential of the collections themselves (Berendsohn 2003b). In any distributed system that includes taxonomic information, there will however, be difficulties in reconciling the varying technical implementations and the complexities of different and changing taxonomic schemes. These problems are of an order that provides a rationale for major initiatives, notably ENHSIN (Scoble 2003), BioCASE (BioCASE secretariat 2002), and the predecessor of BioCASE, BioCISE (Berendsohn 2000). The overall concern, as expressed in the scientific and technical objectives of the BioCASE project, is to remain true to the strengths of taxonomic data, while also realizing the potential that digital solutions offer: "The challenge lies in the provision of adequate standardised metadata, which, on the one hand, do not violate the complex and changing scientific (taxonomy, ecology, palaeontology) and political/historical (geography) concepts involved, but also allow user-friendly access to the information contained in biological collections" (BioCASE secretariat 2002). The solution of "metadata" (a standardized set of descriptors for data, allowing for the development of mechanisms of translation between different databases) claims to allow local solutions to persist while providing for common access portals to be developed (but see Bowker 2000 for some provisos).

We have seen in this section that the ordering of physical specimens in space according to systematic schemes is considered both a strength and a weakness. In introducing collections databases systematics has striven to maintain the acknowledged strengths of existing arrangements, and the wealth of information in the material artifacts, while recognizing and attempting to realize the potential for flexible and distributed access that collection databases offer. The current political climate with respect to biodiversity information provides some of the impetus for funding large-scale experiments in this respect. A preoccupation of such experiments has been to maintain the diversity of locally appropriate solutions already in existence, while allowing for the development of portals for seamless queries across databases. The computerization of collections to provide the content for such systems is also, however, a considerable burden. This burden has been motivation for various experiments to maximize the efficiency of digitization while maintaining integrity of the material in the forms that systematists would expect to use it.

This section therefore demonstrates that the ordering and location of objects traditionally of concern to curators of natural history collections is manifest in a new form with the advent of specimen and collection databases. A particular concern is to develop information systems that, on the one hand, are seen to be true to the material which they purport to represent, and on the other hand visibly meet requirements to make data accessible in user-friendly forms. While new initiatives on an institutional and international scale have come into being, and new roles and sets of skills come into prominence, much of the impact on working practices has been absorbed by curatorial staff. Digital systematics is far from labor-free systematics. The immense burden of integrating legacy systems, including databases and card indexes, means that just as with so many aspects of systematics the payoff in terms of efficiency of current initiatives in databasing will not be visible in the short term. Systematics is accustomed to the idea that material may not be used nor literature be cited for decades or longer. Systematics sees itself as an archival science, whose materials and literature have a persisting relevance. In this context, it is reasonable to see current initiatives as a contribution to an archival project whose payoff will become apparent only on an elongated timescale. With this perspective, we can begin to see how glib it is simply to assume that use of ICTs will automatically bring short-term efficiency gains.

Systematics on Display

Thus far, the discussion of the material culture of systematics has focused largely on the use of collections by systematists. A part of the politics of contemporary systematics, however, focuses on the broader pressure to break down boundaries between publics and scientific experts, and for museums to view themselves as facilitators of learning rather than conveyors of knowledge (Hooper-Greenhill 1992). Hooper-Greenhill shows how the form and meaning of museums has changed with the changing social, cultural, political, and scientific values of the time. In the same way, the form of museums now, including the natural history collections described here, and within that the role of virtual specimens and databases, is going to be shaped by far more than the surface potential of the technologies and will manifest current understandings of the science–public relationship.

The trend toward inviting the public "behind the scenes" takes a very explicit form at the Darwin Centre at the Natural History Museum in London. This Centre, the first phase of which opened in 2001, has a mission to integrate the scientific and display work of the museum. In practice this means a program of tours of the collections, a series of presentations by museum scientists that can be both attended in person and in most cases viewed on the Web site, and an exhibition explaining the work of the museum in collecting, curating, and studying specimens. This exhibition is also available via touch screen at the museum or via the Web site. The work of the Darwin Centre combines actual and virtual display and collections in some interesting and complex ways. The building itself is a carefully designed space: the Darwin Centre Web site contains an extended account of the building's design and its construction process. As described earlier the design is, in the first instance, oriented toward the provision of storage space for the museum's collection of specimens preserved in alcohol. In addition to the practical aspects of design related to specimen storage, the building also illustrates ideas about the working practices of systematists, and about the relationship of their work with the public interest.

A program of tours is advertised as offering behind-the-scenes access to collections for the public. Small prebooked groups are collected from the atrium, equipped with white coats, and led by a member of curating staff through sets of timer-controlled double doors into the specimen storage space. The tour pauses by a set of shelves that has been parted, to show specimens notable for their size, or their age. Proceeding to a different floor, the tour sees that a different arrangement serves for the storage of large specimens, in a room filled with shiny steel tanks, their contents hidden from view by lids liftable by a system of hoists overhead. The walls are lined with shelves which hold large glass jars, many so old that their glass has begun to warp. Passing on to the other side of the building, the tour sees rooms designed for various activities, including parceling up specimens for loan, preparation of specimens, and examination of whale stomach contents. Mostly the rooms are empty of personnel. The tour guide stops and explains what goes on in each of the rooms, but there is little activity to see: he complains that scientists vanish when they hear tours approaching. Although the touch-screen exhibitions make much of the active work that museum scientists do with specimens, the tour does not expand on this aspect. As an experience of what scientific work is like, then, this is a limited portrayal. The focus is on the specimens themselves, their history and the remarkability of their storage conditions. For additional access to the practices of scientists, the "Darwin Centre Live" presentations give a greater, although still thoroughly managed, insight.

Darwin Centre Live is the program of presentations given at the center, largely by museum staff although also by some visiting scientists. Presentations focus on aspects of the work of the museum such as explorations of the biology or history of key specimens, presentations of the process of collecting and curating, and topical issues concerning environment, conservation, or science in the media. Through these talks, the museum is represented as a living repository of expertise, responsive to current concerns and to public interest. Through the presentations themselves, and their re-presentation on the Web, the museum is able to manage a form of accessibility. In achieving this accessibility, the virtual and the material are interwoven, and where one form of accessibility finishes, another takes over.

To illustrate this point, I should describe in more detail the experience of attending a Darwin Centre presentation in the flesh. Here is an excerpt from my field notes:

> I arrive on time, I think, but there is already a fairly large audience seated, and the introduction is underway. First there is a briefing on the setup, and need for microphones when asking questions so that the Web audience can hear them, then two trailers for coming events. The trailers are film, with music, captioned at the end to give the title and date of the presentation.
>
> Once the trailers are over, the presenter visibly shifts into a different mode of presentation, as speaking into his microphone he bids formal welcome to the audience here today and the audience on the Web.
>
> The event is organized as a conversation between the presenter and the expert. The room is set up with four screens across the back. The middle two screens show the slides illustrating the talk, in conventional style (controlled by the expert, who is looking at a

laptop placed in front of him). The outside two screens show what the Web audience is supposed to see, which comprises camera shots of the presenter and expert, depending on which is speaking at the time, interspersed with views of the slides that are being talked about. A camera operator is based at the front, to one side of the central rank of audience. Away to the right-hand end of the room (right as judged when facing the screens) is someone manning what appears to be a mixing desk.

The room itself is long and thin, effectively an alcove off the main Darwin Centre exhibition space. The room is open along one long side, which can be a problem when people come in partway through the presentation and start talking loudly. The audience sit on a couple of ranks of chairs facing the two presenters, and to either side on ranks of benches ranged at angles. These are clearly special benches, made of solid wooden ply, with the ply arranged as a decorative feature. Castors on the benches mean they can be moved to configure the space differently. Inset into the surface of the benches are biological illustrations. Similar illustrations are set into the roof beams. Panels on the wall bear quotations from Charles Darwin. The Darwin Centre logo is projected onto the wall.

The presentation is about Walter Rothschild, eccentric collector who founded the museum at Tring which is now a branch of the Natural History Museum. The focus is on details of the man and his approach to collecting, illustrated by archive photographs of Walter, his family and his museum. At the end, questions are invited from the audience, but none are forthcoming. The presenter assures us that anyone too shy to speak up into the microphone can ask questions later, as the expert will be staying around for a short time. We applaud, and the camera pans round to show us. I am quite surprised to see myself as part of the audience being filmed. As the audience disperses at the end, I realize that many of them do seem to be museum staff. I wonder whether they come along unprompted out of interest, or whether they are encouraged to come along and swell the ranks of the audience.

This Darwin Live presentation prompted me to think about the connection between the Web site and the live experience. The audience at the live event is clearly being offered up as part of the spectacle for the Web audience. We, as the present audience, see ourselves being filmed, and watch the presentation both as it is happening, and that happening as it is being recorded for the Web. We, as the Web audience, get to see some people with whom we are expected to identify as being like us, actually there in the Darwin Centre. We see public access to the museum in action, twice over: in the fact of accessing the archived presentation, and in seeing the audience who were actually at the presentation. At the same time, there are elements of the live experience that members in the Web audience do not have access to, and their attention is drawn to the fact by comments that the presenter makes. At one presentation I attended specimens were passed around the audience and the Web audience could not, of course, participate. This time, the live audience was invited to stay behind and ask questions, in a space not archived for the Web. For much of the presentation the audience is treated as singular, but at moments like these the Web and live audience split. Of course, the Web audience is privileged in other ways, in that they have no need to travel, and gain the ability to replay and fast-forward the presentation, to watch it at a time that suits them.

Museum presentations of natural history specimens have tended to be a "cleaned up" version, showing the end products of scientific theorizing without giving access to the process or opening up questions for debate (Star 1992). It is clear that the kind of behind-the-scenes view that the Darwin Centre gives is also a thoroughly managed access. As Allison-Bunnell (1998) describes, a public portrayal of the work of a museum such as this gives us insight into the ways the institution wants to be seen, rather than direct access to the museum's internal workings. This much is fairly obvious, and by no means a criticism: who would expect a museum to give open and uncontrolled access to working spaces, meetings, tea breaks, and the like? The major natural history institutions have always been heterogeneous spaces, combining public, leisure, and domestic functions with research and teaching sites (Lenoir and Ross 1996; Outram 1996). Managed public access to significant laboratories and large-scale science facilities is fairly routine, as Agar and Gieryn describe (Agar 1998; Gieryn 1998). It might be that precisely this kind of managed interpretation is appreciated by audiences, rather than an unmediated access to the messiness of everyday taxonomic life. What is more interesting for my purposes, however, is to note the ways in which ICTs have been interwoven with on-site activities, such that behind-the-scenes access is achieved across the globe.

The Darwin Centre is a managed forum for describing the work of systematics, arising within a political context that urges public access to this kind of scientific institution. While the Darwin Centre focuses on the processes of systematics and the role of collections within those processes, other aspects of the online work of the museum do similar work with the products of systematic research. Within a context in which there is pressure to be seen to allow the public access to the scientific work of museums, taxonomic and specimen databases on the Internet can serve as a way of responding to this pressure (Natural History Museum 2003). This aspect also becomes apparent with ePIC, the electronic Plant Information Centre of the Royal Botanic Gardens at Kew. This Web portal provides access to the public databases Kew holds, via searches based on the scientific plant name. The project was funded through the Capital Modernisation Fund, a UK government initiative to improve the quality of public service delivery.

In both the Natural History Museum and the Royal Botanic Gardens at Kew, access to databases was initially made available to a general public without adding extensive interpretative materials or guides aimed at making the information accessible to a lay audience. The information that has been made available has largely been created according to other sets of priorities. Some exceptions to this at the Natural History Museum relate to the recognition of the historical interest value of certain collections and the use of the Web as a supplement or alternative to visitor exhibitions at the museum. The pressure for behind-the-scenes access has, however, in the case of online databases often been interpreted as meaning access to the raw databases that might also be used by systematists. While the Darwin Centre employs virtual technologies and real tours to provide a very managed form of behind-the-scenes access, the online databases of the Natural History Museum and ePIC provide a less managed and more direct access to the work of systematists past and present.

The provision of access to collections databases via the World Wide Web is a phenomenon not confined to systematics collections. Museums more generally are subject to pressures to improve accessibility, and in the current climate this often translates into the use of digital technologies. In the UK, the Designation Challenge Fund, operated by the Department of Culture, Media and Sport, offers competitive funding to recognized museums for "imaginative and innovative schemes to boost access" (Resource: the Council for Museums Archives and Libraries 2002). Of the forty-nine grants announced in July 2002, thirty made explicit mention of ICTs. These entailed either provision of online access to collection databases or the development of interpretation and learning materials to be made available via the World Wide Web (or in a few cases both). Programs of digitization, documentation via database, and imaging of objects are frequently mentioned. The overall impression is that the museum community and its funding bodies have embraced ICTs as a means to provide public access both to collections and to learning experiences. Again, we find a mixture of interpretation and raw access. Both collections databases, and online learning experiences with little or no dependence on collections of physical objects, find a place on contemporary museum Web sites, and are increasingly an expectation of a museum that is meeting the obligations placed on it by receipt of public funding.

Systematics collections, often housed within major museums, have become a part of the trend toward ICTs as a means of promoting access. Initiatives at the Darwin Centre thus need to be seen as a part of this wider picture. Systematics inhabits two distinct spheres of politics, funding and working practices, as a part of the growing international consciousness of the need to conserve biodiversity and as a component of the museums community. In each sphere there is a drive toward the digitization of collections and the provision of electronic access to information. Given these two sets of pressures, the developments in use of ICTs within the material culture of systematics that I have discussed make a lot of sense, although their introduction has not happened without tension, struggle, and compromise. In the final section of this chapter I will draw together implications for policy hopes of transforming science through ICTs. I will also explore constraints on some of the more radical applications of ICTs.

The Ambiguities of Digitization

Systematics, the science that classifies organisms and studies relationships between them, has a distinctive material culture. Objects are collected, stored, ordered, and examined via practices that constitute them, collectively, as an instrument for studying the diversity of the natural world. This common goal takes different guises across the discipline in relation to the varied characteristics of different groups of organisms and the ways in which systematists work with them. The uses that are made of ICTs in this context are aimed at preserving the integrity of the instrument and increasing its usability and availability. Systematists themselves have argued for their working environment to be seen in this light. Wheeler links current developments in cyberinfrastructure with systematics in this regard: "Cyberinfrastructure has the potential to unite the world's museums and taxonomists into a seamless virtual biodiversity observatory. This observatory, like a telescope, would allow scientists to see biodiversity from new taxonomic perspectives. . . . This taxonomic platform would be much more than a virtual biodiversity observatory, however, and constitute the most powerful instrument ever conceived for taxonomic research and education" (Wheeler 2004: 579). Historically, the material culture that constitutes the instrument of taxonomy has been distributed among spatially dispersed collections. The integrity of the system as an instrument for viewing natural diversity has been maintained by the mobility of taxonomists themselves and by the loan system which sends individual objects for study elsewhere. Through these practices taxonomists have created their individual instruments out of the metainstrument. Taxonomy has, in effect, been maintained as a space of flows (Castells 2000) long before ICTs came along. The instrument of taxonomy is maintained by flows of people and objects between the nodes of the systematics institutions. Each of the nodes has some commitment to its location: the collection is often biased toward particular geographical specializations thanks to location or history. Nodes, however, are also detached from their locations, in that they are part of the wider taxonomic system.

ICTs, then, are being used to substitute for some of the movement that already goes on in this space of flows. Accompanying this has been a process of specifying the valued qualities of objects for which it is felt that ICTs cannot substitute. ICTs are complementing the existing flows, helping to maintain and enhance geographically dispersed taxonomic collections as part of a metainstrument for exploring the diversity of natural organisms. At the same time, in the context of current pressures on taxonomy, ICTs are being deployed to make this instrument accessible to wider audiences. Sometimes, as in the Darwin Centre, this means attempting to display the inner workings of the instrument. In other cases, it means making the end products more readily accessible, with the hope of making current activities more efficient or overcoming geographic inequities.

The virtual culture of systematics thus appears, in many ways, to be continuous with its material culture. The concerns about preserving specimens, working with them, and using them as the basis of comparative analysis transfer from the material to the virtual. In some cases the virtual specimen appears to resolve problems with physical objects; on other occasions the virtual specimen seems not immediately to meet the needs of the existing practices, and other solutions have to be pursued. The virtual culture of systematics is not necessarily a radical imposition on existing practices, nor is it a scene of outright transformation; but it can be a relatively gentle reworking of old practices through new forms. Even within systematics, new technologies are not always presented as radically novel. Godfray and Knapp make another form of link between new practices and old, by citing a fundamental continuity with the ways that systematics has traditionally proceeded: "For example, taxonomists should continue to embrace the use of the Internet and related media to enhance taxonomy, and fuse modern molecular approaches with traditional morphology. This is what Linnaeus and the great polymaths of the eighteenth century would have done if alive today" (Godfray and Knapp 2004: 569). Just as Myerson (1998) argues for the electronic archive in academia more generally, systematics seems to alternate between visions of a new virtual culture that transforms practice and a view of the virtual as the continuation or even concretization of traditional working forms. Possibly, then, it is mistaken to pose the question as a choice between virtual culture as either continuing or transforming traditional systematics. This chapter has shown the complex ways in which the two are interwoven, such that it begins to seem unfair to make a simple choice. Virtual cultures are ambiguous cultures. For all the times that we think things or procedures have been transformed out of all recognition, there will also be occasions when we feel that nothing has changed at all (Woolgar 2002b). Myerson's (1998) depiction of the ambiguity of the electronic archive resonates with the experience of systematics with virtual technologies.

In chapter 2 I discussed the policy hopes that surround the use of electronic technologies. New ICTs were seen as ways of promoting efficiency and effectiveness in scientific work, encouraging sharing while discouraging duplication of effort. ICTs were seen as ways to change scientific culture, as a policy lever that could be applied without needing to make specific demands for behavioral change. So, given the analysis presented in this chapter, would ICTs work as a lever to change systematics? The answer is that they could, to some extent, but this lever could move far more than was ever intended. The extent to which ICTs have acted (from a policy perspective) as facilitators of desirable change in systematics is debatable, but there clearly have been some important innovations, which revolve around the co-construction of material and digital specimens and the practices for working with them. The notion of the digitized collection as an accessible, enumerated, and more easily curated collection is important in this regard. Paradoxically, it is as an intensely material culture that systematics has been most susceptible to political pressures to virtualize. Its very materiality makes it prone to characterization as inaccessible, aligning neatly with the prevailing rhetoric that to be online is to be accessible. The physical collections stand to be more thoroughly known and effectively used through their preservation in digital form. Similarly, the belief in online information as accessible information would seem to mean that the advances in online presentation of taxonomies, identification keys, and interoperable databases of taxonomically linked information cannot help but advance systematics, and biology more broadly, toward a more efficient and more effective way of working.

It would appear, then, that the encouragement of ICT use could act as a lever for change in systematics. However, there are some important limits on this leverage. Among these limits are the extent to which digitized collections are tied into existing working practices and institutions. The full radical potential of digital technologies is unlikely to be realized under existing conditions of funding, institutional politics, and disciplinary culture. A culture is not obliged to change itself radically in the face of a technological innovation, when so many other influences on it remain unchanged. The leverage applied by a new technology may also not be entirely what policy either wants or expects. Within systematics, the emphasis on digitization as an intrinsically desirable thing means that not all decisions to digitize receive careful and critical attention. A blanket imperative to digitize can mean that systems are developed in the hope, rather than the firm expectation, that they will find use. A general leverage across a whole field can move far more than was ever intended. It would be interesting to see the effects on the field that would result from specifically making funding available for users of taxonomy to identify themselves and get involved in shaping developments.

As it stands, institutions' response to the pressure to participate in digitization initiatives is mediated by other concerns they face, including the need to maintain a distinctive institutional identity and protect their unique assets in the form of specimen collections. In chapter 6 I consider the structuring of the disciplinary field that results from the attempts of individuals, institutions, and initiatives to navigate these pressures. Before moving to this structural mapping, however, it is necessary to look at another important structuring influence on the possible responses to pressures to digitize: the discipline itself. Thus far the discipline has been taken as a self-evident backdrop to digitization initiatives. In the next chapter I look at the discipline as constituted in and through particular ways of communicating. This view of the discipline as a dynamically constituted construct mediated through specific communication practices casts further light on the role of ICTs as sites for imagining science.

5 Communication and Disciplinarity

The subject of communication has already arisen several times in my discussion of the way that contemporary systematics has adopted ICTs. As explained in chapter 2, science is reliant on a variety of more or less formalized communication mechanisms to enable its practitioners to work together and learn from one another, and to allow scientific knowledge to have a public presence outside its place of production. ICTs have, as chapter 3 showed, offered a provocation for thinking about ways that systematics could communicate better, but these predictions of improved possible futures have not always been thought through in terms of what is culturally sustainable for the discipline. In this chapter I will be turning the focus more directly onto the ways that systematics conceives of its communication practices and evaluates their effectiveness. I will be particularly concerned with the ways in which the discipline is constituted in and through diverse means of communication. Aspects of taxonomic communication that were taken for granted in the previous chapter, such as the publication of monographs or the use of the Taxacom list, here become the focus of specific attention. I will consider how these ways of communicating become commonplace, what meaning they have for their users, and to what extent new communication media have offered viable new possibilities for systematics. Communication practice proves to be an interesting place to look for the mechanisms through which change may be achieved. In chapter 4, looking at the material culture of systematics focused attention on specimen collections hosted by institutions, and few resources were found that crossed institutional boundaries. In this chapter exploring communication practice shifts our attention to the discipline. Both by acting as the arbiter on appropriate means of communicating and by being enacted through various forms of communication, the discipline acts as a lens through which the dynamics of change and continuity in communication make sense.

The chapter divides into four sections. The first introduces the context in which communication regimes in science develop, mapping out an expectation both that regimes will have disciplinary specificity and that there will be mutual visibility across disciplines. This forms a basis for the next section, which looks at ICTs in relation to the formal communication systems of systematics. The potential for change turns out to be articulated through a development of rationales for what the discipline needs and portrayals of the appropriate kinds of change for the discipline to undergo. The next section moves to informal communications, looking at the role of ICTs in the form of a mailing list in constituting the discipline. The final section then gathers these observations together to consider how far systematics represents a special case in its response to ICTs as potential facilitators of new communication regimes.

Imagining Change in Communication Regimes

In chapter 3 I described the "Godfray vision" for the transformation of systematics through a new genre of communication, which would deploy Web sites to publish a consensus classification for each group of organisms. These Web sites would intervene in the existing communication system of systematics in profound ways, altering the relationship of the taxonomic community to its products and its users and diminishing the status of prior literature in the practices of the discipline. Godfray's vision was inevitably controversial to systematists in its wholesale revamping of existing practices, yet it was appealing to a wide range of commentators for its focus on user needs and its deployment of the latest technology. It stimulated much debate around publishing practices and the state of the discipline. Chapter 3 described the role of the Godfray vision as an occasion for strategic portrayals of the discipline and its progress with new technologies. In this chapter I focus on the existing publishing system in systematics, and explore the ways in which portrayals of this current practice have been used to shape the possibilities for innovation. It will become clear that Godfray's vision, though provocative, was ultimately unlikely to succeed in the face of existing priorities and the prevailing arrangements for the sanctioning of change. The imaginings of potential for change relied on particular portrayals of the current system and of the heritage of the discipline. Just as in the previous chapter we saw that virtual culture provided new occasions for imagining material culture, so in this chapter we find that new digital communication possibilities become opportunities for mapping out particular versions of their antecedents.

As we saw in chapter 2, communication is constitutive of the scientific endeavor, and yet the diversity of disciplines means that each may operationalize its communication in a quite different way. Chapter 2 introduced the idea of computer-based communications as situated phenomena that are brought into being and given meaning in diverse social contexts, and also give rise to new but similarly diverse contexts in use. Together, these two sets of observations imply that we should look to systematics to find some specific ways of deploying new communication media in the work of the discipline. While the prevailing cultural expectations may lead systematists to use these media in ways we will recognize from other settings, it is also likely that we will find specificities unique to systematics, and that we will find the contexts that systematists develop will be particular to their concerns.

It is thus important to ground this discussion from the outset in an expectation of specificity. It is also, however, crucial to recognize that the future of scientific communications has been a topic of considerable attention, animated by a range of stakeholders (Kling, Fortuna, and King 2001; Kling, Spector, and Fortuna 2004). Discussions about future publishing have attracted concern from publishers, from scholarly societies, and from funding bodies (as in, for example, recent Wellcome Trust reports (Wellcome Trust 2003, 2004). The scale of the attention is captured in a recent bibliography, containing over 1,300 publications focusing on the topic of e-prints and open access journals (Bailey 2005). It is therefore important to temper expectations of specificity with the recognition that changes to communication structures are at once within the scope of disciplines and influenced by a range of other actors. If a discipline steps outside of prevailing expectations to any great extent it will probably need to justify its decisions as a special case. In the descriptions of systematics that follow it is again important to remember that systematists are often conscious that they are speaking to a wider audience than the members of their own discipline. When they call for the use or rejection of new technologies they are also aware that the general climate views these technologies favorably.

In addition to the role of influential actors, the will to transform scientific publishing is then also influenced by the availability of new technologies and the cultural connotations those technologies come to carry. The Berlin Declaration places the agency within the technology itself, noting that "The Internet has fundamentally changed the practical and economic realities of distributing scientific knowledge and cultural heritage" (Conference on Open Access to Knowledge in the Sciences and Humanities 2003). It might more subtly be argued that expectations about the Internet have allowed for ideas about transformed scientific publishing to be propagated. Noting that Internet communication has become commonplace within society more broadly suggests that it will become increasingly difficult to argue against its use in scientific circles, and that the open access movement will both gain momentum from and help to sustain those wider cultural currents. These ideas will, however, be consumed in different ways within different disciplines, and the technology in itself does not guarantee the form of the outcome. Scientific communication systems are not, after all, simply the upshot of a particular technology, but are sociotechnical ensembles which combine a technology with social arrangements for how it should be used, such as provisions to encourage authors to participate and institutional, legal, and economic arrangements to sustain its operations. These communication regimes (Hilgartner 1995) produce sociotechnical ensembles that are quite specific to different disciplinary situations. As Kling et al. argue (2004), it can be difficult to institute a new communication regime that differs dramatically from existing arrangements.

Taking these observations to the specific case of systematics, the prevailing assumption described in chapter 3—that systematics needed to modernize, and that the use of ICTs was a route to being modern and to being seen as modern—suffuses the specific choices about communication media that are discussed in this chapter. We might therefore expect to see a willingness to develop new communication regimes to exploit the practical and symbolic qualities of these technologies. Also, as we saw in chapter 4, the taxonomic endeavor has traditionally been widely distributed across geographic space, and we might therefore also expect to see a willingness to use computer-mediated communications to support existing distributed activities and develop new ones. In addition, we might expect to see a willingness to explore open access models, thanks to the relatively low influence of commercial publishing houses. Commercial publishers have always been less important in systematics than in some other scientific fields, since the highly specialist nature of publications makes them relatively unattractive as commercial concerns. The resistance noted from these stakeholders to open access publishing in other disciplines might be less evident here.

Against these factors promoting systematics as a site for new forms of communication is ranged the history of the discipline. Systematics has an unusual and highly formalized approach to publication, which makes it difficult for new initiatives in formal publications using new media to become established at a grassroots level. In the next section of this chapter

I will describe the established conventions of taxonomic publishing in order to provide the backdrop for a discussion of the various factors that shape the appropriation of computer-mediated communication for the formal publishing activities of systematics. We find a considerable tension operating between the desire to embrace the new and the concern to preserve valued qualities of the old, and between evocations of the Internet as the technology of the future and fears of its ephemerality. In this description I discuss the medium as shaped through the concerns of the discipline. In the second half of the chapter I turn to the informal communications of systematics and a consideration of the medium as constitutive of the discipline's sense of itself.

There is a tendency, which I have manifested in previous chapters and in the introduction to this chapter so far, to talk about scientific disciplines as if they have an inherent integrity that preexists particular practices. This is misleading, to the extent that the discipline is constituted in and through its practices, and any sense of the discipline as a discrete entity is available only as it is constantly reasserted through various means of communication. We recognize a discipline, and its practitioners recognize it as such, through the scholarly societies, conferences, journals, textbooks, and newsletters by means of which members tell one another what the discipline is. Members develop their sense of what the discipline is through the various representations available to them. It is this thinking that informs the consideration of the Taxacom list, in the second half of this chapter. I aim to show that it is misguided to see communications media as shaped solely by the disciplines that use them: rather, disciplines and media turn out to be co-constitutive.

Systematics as a Culture of Communication

Chapter 4 contained some hints about the particularities of taxonomic publishing. One of the issues that specimen collection databases raised was that annotations on specimens might, if published in databases, place in the public domain names that had not been formally published. There is a formal mechanism for the publication of new names, designed to make sure that their application is unambiguous and that the information remains accessible to the community. This formal mechanism is, however, only a small part of the overall picture of communication in systematics. Formal nomenclatural codes take a place within a rich communication culture, which combines different media and genres, and builds on conventions developed over centuries to support the practices of systematists, sustained both through the coordinated activities of major institutions and the day-to-day practices of individual specialists.

The picture of the systematics communication system that I give here is framed in general terms. In arriving at this picture I draw on a more detailed study of communication among systematists focusing on the Collembola[1] which I conducted previously (Hine 1991). I reviewed the literature reporting on the taxonomy of the Collembola for the thirty-year period from 1959 to 1988, taking an interest in where taxonomic revisions were published, how often, and by whom. I also undertook a citation analysis of this literature and made a tentative attempt to discover what factors affected the acceptance and formal acknowledgment of taxonomic revisions. I will not repeat the details of this study here, but will draw on observations about the qualities of the taxonomic literature of the Collembola to illustrate a more general picture of the communicative culture of contemporary systematics. In mapping out what this culture is like I will be making explicit something that is, apparently, an accepted part of the expertise of a taxonomist. They need to know where information is to be found, and thus they see their work with the literature as an integral part of their role: "The taxonomist will know intimately the information landscape for his or her particular group, and in fact this is as much a part of 'knowing' a group as being able to identify specimens" (Godfray and Knapp 2004: 563). This chapter offers a broad sketch of the kind of territory that this intimate knowledge encompasses.

Systematics builds its communicative culture on the existing system of refereed scientific journals. Much novel taxonomic work, including descriptions of new species, is made available through submission of papers for publication by journals. Other outlets for systematic work include book-length studies which either report on the taxonomy of a group of organisms or provide a checklist of the organisms to be found in a particular country or region. These large-scale synthetic works tend to be produced over a long period of time, may run into several volumes, and may represent one's "life's work" for the systematist(s) concerned. There are many acknowledged gaps where a synthetic work is lacking. In these cases, an overview of the taxonomy of a group can be formed only by consulting a widely dispersed literature in a variety of journals. In the Collembola, for example, I found the taxonomic literature for the period that I studied spread across 122 journals and books. Whereas core journals could be identified that focused on soil ecology or represented the institutional journals of key workers in the field, there was a long tail of journals that contained only one or two relevant articles. Of course, a given taxonomist may publish in relatively few journals, but the net result is a widely dispersed primary literature. Print runs are often small for these specialist titles, and they can consequently be hard to obtain outside the few major specialist libraries.

Formal literature may often be supplemented by informal sources. In the Collembola an informal newsletter was circulated for many years to interested parties, keeping them informed of new literature. My choice of this group to focus on for a taxonomic literature analysis was due in part to the fact that my supervisor had a full back archive of this newsletter, and I therefore did not need to begin a search of the literature from scratch. Although I had no means of judging that my coverage was comprehensive, I could at least be clear that I was using a route into the literature not unlike that available to collembologists, and that literature invisible to the newsletter was effectively invisible to that community. In the past, this newsletter for the network of experts interested in the Collembola circulated as a set of duplicated notes. More recently, a Web site (Bellinger, Christiansen, and Janssens 1996) provides access to indexes of names, identification keys, experts in the field, literature, and images. The breadth of provision is vastly beyond anything that the paper newsletter could provide. The ethos, however, is somewhat similar. The site is maintained freely by its authors and the contributions of many members of the community to its resources are acknowledged. In a more similar style to the original Collembola newsletter, the Botanical Electronic News (http://www.ou.edu/cas/botany-micro/ben/) operates as a volunteer-run electronic newsletter available both via mailing list subscription and Web-based archive.

In addition to this kind of community-based resource, more formal indexing sources are also important in taxonomy. So complex and widely spread is the taxonomic literature, and yet so rigid the insistence on the relevance of past literature, that the field has produced a rich array of indexing and notification services. For the animal kingdom, *Zoological Record* (published by BIOSIS until 2003 and subsequently transferred to Thomson Scientific) acts as an index to the literature on all fields of animal biology, in the process acting as "the world's unofficial register of animal names" (BIOSIS n.d.). The Royal Botanic Gardens at Kew publish the *Kew Record of Taxonomic Literature* (Royal Botanic Gardens Kew 2000), among other indexing and nomenclatural services. In recent years databases have become an acknowledged solution for making the indexes previously compiled on card file or in bound volumes more widely available. The heritage of indexing and abstracting services, from small communities, large institutions, and publishing companies, provides a basis for more recent initiatives to make information available in electronic form and on the Internet. This aspect of the virtual culture of systematics shows strong continuity with the communication culture of the past.

Large-scale indexing projects have lately been linked with the idea that taxonomic information needs to be gathered together and rationalized, to improve stability of names and meet the expectations of users. Lughadha (2004) addressed the question of creating digital nomenclatural resources in a discussion of the prospects for an electronic list of all known plant species. Lughadha describes the development of the International Plant Names Index as a joint venture between Kew, the Harvard University Herbaria, and the Australian National Herbarium. This digitally accessible resource, created by merging existing resources, aims to replace the need in most cases to consult the primary literature for nomenclatural information. Merging data sets, which each contain incomplete information and may also conflict with one another, and keeping the index up to date for new nomenclature and newly discovered but previously unindexed literature, is a massive task. The approach taken by IPNI has been to promote contributions from the user community. Recently this has involved a mechanism by which contributions can be made automatically through the Web site, immediately made visible with a "pending" flag, and subsequently checked by an editor who will consider granting approved status. This kind of contribution might be made by a taxonomist who, making use of IPNI to retrieve names for a group that they are working on, finds missing or erroneous data. The IPNI solution is designed to allow an immediacy of access while retaining an editorial control, thus addressing concerns over data quality from the two different angles of up-to-date coverage and authoritative treatment. Virtual availability of an indexing service here offers the possibility of a new form of connection between taxonomists and the information provided, although it remains to be seen how widespread the use of feedback facilities will be.

Indexing services such as IPNI are important in the context of the very particular relation of taxonomy to its literature. This relationship is enshrined in the codes of nomenclature, which describe the process by which names are considered to be effectively published and the steps to be taken when more than one name appears to be available for a given taxon (see Winston 1999 for a description of the history of the codes and outline of their provisions; also Jeffrey 1989). Separate codes exist for botany and zoology and for viruses, bacteria, and cultivated plants. The development of nomenclatural codes began in the nineteenth century, and though the original aim was to produce a single code, disagreements soon produced a split between botanists and zoologists. Subsequent developments have produced systems that remain very similar and operate in parallel, each being accepted discipline-wide as enshrining required practice. The botanical code is maintained by the International Association for Plant Taxonomy, and the zoological code by the International Commission on Zoological Nomenclature. In each case there is provision for appeals to deviate from the code in order to preserve the stability of a particular name.

The codes were initially designed to promote the stability of names by giving clear grounds for deciding between competing synonyms. The long-term relevance of the taxonomic literature is based on the principle of priority used within the nomenclatural codes, according to which the first published name for a given taxon takes precedence over any subsequent names. It is thus possible for a name in common use to be overturned if an earlier publication is found that gives the same taxon a different name. Publications remain relevant for this purpose back to a nominated starting point, which for plants can be as early as 1753, and for animals is 1758. Godfray was certainly not the first to propose that there should be a new starting point for priority purposes. There have been previous occasions when more recent starting dates for botany and zoology have been proposed, and there have been attempts to institute systems where lists of names in current use would form a new starting point. There have also been proposals that publication could be made more transparent if new names were to be registered with a central index in order to be considered validly published. For bacteria, a new starting date for nomenclatural priority and a system of registration have been instituted and found acceptable thanks to particular problems of interpreting past literature in this field. For botany and zoology these solutions have met with some resistance among those who see them as making taxonomy more likely to be dominated by elite groups. Any changes to the codes must be voted on, and radical changes, such as new priority starting dates, name registration, or indeed Web-based taxonomy, seem bound to meet with considerable resistance.[2] Any apparently centralized system will be politically unpopular within the current systematics community acutely aware of existing geographic imbalances and concentrations of influence.

In contrast to the suspicion that faces any proposed radical transformation of nomenclatural codes, there has been a considerable interest within systematics in digitizing the legacy literature. Whereas in many fields of science where publications have a shorter half-life this would not be a priority, in systematics the long-term relevance of the published literature makes digitization of libraries a sensible task to consider. This makes sense both in terms of the political focus on sharing of taxonomic expertise, and in the context of interest in improving the image of systematics as efficient and focused on the task at hand. Godfray (2002a) accused systematists of being distracted by searches in obscure literature. His proposed solution was to start again and render that literature irrelevant for nomenclatural purposes; a more culturally viable solution for systematists seems to be to retain the significance of the literature but use digitization to make it more accessible. Large-scale digitization of legacy literature is also easier to contemplate in taxonomy than in many other fields, since copyright concerns are limited, both by the low involvement of commercial publishing and the sheer age of much of the literature.

Acknowledgment of the need to digitize taxonomic literature is becoming widespread. The first international conference of the Entomology Libraries and Information Network, held at the Natural History Museum in 2000, was entitled "Insect Information: From Linnaeus to the Internet," and included presentations on the difficulties of accessing entomological literature in less developed countries and the advantages of digitized literature (Entomology Libraries and Information Network 2000). More recently, a conference also at the Natural History Museum in London entitled "Library and Laboratory: The Marriage of Research, Data and Taxonomic Literature" endorsed "the need to make the massive and vital resource of taxonomic literature accessible on line as a major part of the digital workplace for science" (Consortium for the Barcode of Life 2005). The possibility arises of blurring the boundaries between different forms of data, in particular bringing literature and specimen collections closer together by presenting both as database records relating to a given species. In this formulation the distributed taxonomic instrument represented by specimen collections is augmented by additional forms of information, extending not just in space but across previously separate data sources and media genres. As we saw in chapter 3, digital information is construed as available information, but here the important added twist is that once data is digitally available it is theoretically straightforward (although far from easy to achieve practically) for different types of data to be linked together. Instead of being a separate domain, literature becomes part of a seamless sphere of information. This vision attempts to mobilize support for a form of open access tailored to the expectations of the systematics community.

Although it has become popular, however, to lobby for the digitization of the taxonomic legacy literature, current publishing practices have so far remained very little changed. Private databases are often used by taxonomists to organize a project, yet the end result will still often take the form of a printed monograph or checklist. Heywood (2001) talks about the pressures that are pushing away from life's work monographs and toward continually updated Web publishing, yet electronic products like the Flora of North America (http://www.fna.org/FNA/) are still very much the exception. Another notable early exception was Flora Online, begun in 1987 to distribute machine readable botanical information and software to subscribers on disc, and issued until 1993. Although the archive Web site for Flora Online (Zander 2003) notes that it was superseded by the World Wide Web, in practice Web publishing is still not an automatic choice.

The ironies of this situation have been noted, for example in a review of a recently published comprehensive checklist of chewing lice, prepared using a database but published only in printed form: "It borderlines on madness that while some members of the systematic community are busy digitizing taxonomic works produced before the dawn of the personal computer, others are using electronic techniques to publishing new paper based catalogues that will at some future point need to be redigitized!" (Smith 2004: 668). The reasons for the continued emphasis on printed products remain unclear, but are likely to depend in part on suspicions of the ephemerality of the electronic form. Systematists trust print and paper, for they still routinely use books printed decades and centuries ago in their daily work. They like electronic resources as a handy tool, but have not yet invested in them the same degree of trust. It is less likely that systematists are fearful to publish electronically for reasons of credit. In other fields scientists might be concerned that electronic publishing is less prestigious, or less well embedded into the citation system. Taxonomic publishing has, however, rarely operated according to this kind of judgment. Despite building their communications on the conventional scientific system, systematists note some significant differences, particularly in citation practices. There have been complaints that use of citation counts to assess the impact of scientific works is misleading in the case of systematics, with some practitioners commenting that this leads to underestimation of the worth of systematics (Valdecasas, Castroviejo, and Marcus 2000). Many problems have, of course, been noted with the use of citation counts to assess impact. The problem from the point of view of systematists is that the taxonomic literature is rarely cited directly. Many publications are cited only when the next revision of the group is published, which may be years or decades later. Whereas in other sciences a published paper may be viewed as relevant to current work for only a short time span, in systematics the relevance of a paper relates to the work of the individual and state of taxonomy within a particular group of organisms and not to a discipline-wide "cutting edge." If you are working on a revision of a group, any literature on that group going back to the work of Linnaeus in the eighteenth century is potentially relevant.

Although citation of literature is rarely used in systematics, there is considerable emphasis on giving due credit for contributions and on individual reputations. The work of past taxonomists is routinely acknowledged by the practice of citing the authority for names: for example, a common daisy as described by Linnaeus would be formally cited as *Bellis perennis L.* and a form of this daisy described by D. C. McClintock cited thus: *Bellis perennis* L. f. *discoidea* D. C. McClint. This practice, of stating the name of the person who originally described the taxon along with the name of the taxon, is seen as serving to acknowledge their work and also to reduce ambiguity in the usage of names. This dual focus on due credit and on decreasing ambiguity transfers through from naming practices to database initiatives, where records are often marked with the name of the scrutineer or taxonomic coordinator who checked the data. In chapter 6 I discuss further this issue of individual credit for database activities, but here it is important to note that again the established communication culture and the emerging virtual culture are aligned in this regard.

The communication culture of systematics, at least as far as it relates to basic taxonomic activities of describing and naming groupings of organisms, is very formalized. In addition to their focus on the rules of priority, the codes of nomenclature also specify criteria by which names may be considered validly published. The intention of legislating in this way is to make sure that authors of new names take care to make their work publicly available in a durable fashion, such that a future taxonomist would be able to consult the work. The nomenclatural codes view taxonomic literature as an archive of past works to be consulted, and the criteria for publication to be considered valid are intended to ensure the integrity and accessibility of the archive. Further aspects of the nomenclatural codes, such as the priority rule under which determination of the appropriate name for a taxon is assigned to the first valid publication, depend on this initial definition of what it takes for a taxon name to be published.

There has been considerable debate over the way in which electronic publishing possibilities should be treated for these purposes. In the 2000 revision of the International Code for Zoological Nomenclature, the possibility was recognized for electronically published names to be considered valid, in a response to considerable amounts of pressure. However, the prospect that electronic publication might be ephemeral was very troubling in the context of a discipline that considers literature to be worthy of consultation going back more than two centuries. The decision to allow electronic publication was therefore enacted with some considerable caveats and the addition of safeguards to provide durability. In particular, identical copies of the electronic publication were to be made and placed in five libraries. The authors of the code were troubled that many electronic publications are amenable to continuous update, thus threatening the notion of the scientific literature as a true archive. The code therefore explicitly requires that for valid publication there should be numerous copies, and these should be identical and durable. Whatever general features electronic publishing might have to offer, these are articulated through the perceived demands of the particular situation of publishing taxonomic descriptions. There has been considerable enthusiasm to seize the possibilities of electronic publishing, and the open access movement has certainly had influence within this community, but as yet the concerns about ephemerality take precedence and electronic literature remains a secondary source for nomenclatural purposes.

Within the botanical community there has also been considerable discussion of the possibility of electronic publishing of new names. A committee established by the International Botanical Congress was charged with considering the following as key issues for the acceptability of electronic publishing:

1. How can an electronic publication be archived in a form that will remain readable by future hardware and software?
2. How can an original version be protected from changes?
3. How can each publication be uniquely identified for access purposes? (Zander 2004)

Despite most of the committee who reported in 2004 being largely in favor of electronic publishing they did not recommend a single technical solution. Zander attributes this divergence to the committee's representing "not a random subsample of systematists, but a group with strong and disparate visions" (Zander 2004: 592). The proposals voted on at the International Botanical Congress in 2005 included the possibility of publishing new names on the Web and on CD-ROM or DVD, provided that steps were taken to make clear that the publication was intended as part of the enduring scientific record and registered as such (for example via ISSN), and had an unambiguous date of publication. Amended versions of the proposals were passed, and a Special Committee on Electronic Publication convened, to report back in 2011 (McNeill et al. 2005).

The nomenclatural codes have therefore articulated a conception of electronic publishing for purposes of publishing names. Codes, however, like any legalistic form, are notoriously open to interpretation. The once articulated version of electronic publishing expressed in codes of nomenclature is then articulated again through the particular circumstances of an individual's publishing practices or an electronic outlet's attempts to meet the requirements while holding true to their own version of the qualities of electronic publishing. This doubled articulation is likely to produce a variety of genres and of relationships between Web-based publishing and the formal requirements of nomenclatural codes. I will therefore outline some specific ways in which electronic journals have thus far responded to the requirements placed on them by the nomenclatural codes (since botany has only recently accepted the possibility of electronic publishing, both examples are from zoology).

Inevitably some discussions about the prospects and promises of electronic publishing as the way forward for systematics arose in the issue of *Philosophical Transactions of the Royal Society (B)* organized by Godfray and Knapp (2004) under the title "Taxonomy for the Twenty-first Century." Godfray (2002a,c) having become notorious for his suggestions that taxonomy should become a Web-based science, it was unsurprising that contributors to the volume addressed the prospects for change. Knapp et al. (2004) looked at the future prospects for nomenclature in the light of previous developments in nomenclatural codes and practices. They viewed nomenclature as a vital and yet evolving set of practices, but saw it as inevitable that new technologies would have an impact only gradually. The concerns about the accessibility of taxonomic literature over long periods of time warranted caution over embrace of new and potentially ephemeral electronic technologies. They cited ZooTaxa (http://www.mapress.com/zootaxa/) as an example of an electronic publication that has managed to address the requirements of the International Code of Zoological Nomenclature: "The journal is published online, with separates lodged in public libraries to satisfy the requirements of the Zoological Code. The speed with which new taxa are rigorously peer reviewed and published in this journal is astonishing: a look at recent issues shows receipt dates of only a week or so before publication dates; market forces will surely stimulate more such efforts in the very near future" (Knapp et al. 2004: 616). The journal's Web site stresses the advantages of the electronic format for taxonomic purposes, including speed as mentioned in the quotation above, but also the lack of restrictions on length of contribution and color illustrations, flexibility of publishing formats, and the environmental soundness of electronic publishing, coupled with a print-on-demand approach to the print version. Peer-reviewing of contributions is also prominently mentioned, stressing the need for a continuity of quality control while advocating other kinds of change. Adoption of electronic communications is again in this case a question of managing the co-construction of past and future so as to present the future as an appropriate response to both past experience and technological promise.

Palaeontologia Electronica (http://palaeo-electronica.org/) has also met the requirements for Web-based publication, being hailed as the first site to comply with the requirements of the new version of the International Code of Zoological Nomenclature in 2000 (Gee 2000). In the first issue of the journal, the editors take pains to stress the continuity of the new format with established conventions: "It is widely recognized that there is no substantive difference between electronic and print publishing in terms of most scientific publishing conventions (e.g., peer-review, editorial oversight, copy-editing)" (MacLeod and Patterson 1998). Writing before the decision to allow for limited electronic publishing of new names had been made, they depict a full electronic publishing system as an inevitable future: "It is true that the current ICZN and ICBN rules disallow the naming of new species in electronic media. Nevertheless, given the popularity of the emerging electronic publishing industry, it seems inconceivable that this rule will be retained once the electronic publication archiving issue has been addressed" (ibid.). The particular advantages they note for palaeontology include increased graphical possibilities and decreased reliance on a diminishing range of established journals that accept taxonomic works. The editorial in the 2000 issue containing the journal's first paper publishing new names offered a detailed account of the ways in which the journal met the International Commission for Zoological Nomenclatures requirements: "Thus, new taxonomic names or taxonomic actions distributed electronically can be validated provided that permanent copies of the work are placed in at least five named repositories and also made available to the scientific community. We will meet this requirement by pressing and distributing a sufficient quantity of an exact copy of the WWW version of **PE** on compact disks (CD) in archive-quality, read-only format" (Patterson 2000). Whereas ZooTaxa opted for a hybrid electronic and print-on-demand format, Palaeontologia Electronica remains resolutely electronic. Both accept the need to comply with the requirements of the nomenclatural code, but each embodies a different understanding of the relationship between print and Web and of each with the practices of readers and librarians. Both remain unusual in a publishing culture still largely focused on print.

Both ZooTaxa and Palaeontologia Electronica are keen to point out continuity with other forms of publishing in terms of quality control via peer reviewing. There is a prevailing concern expressed in discussions of electronic publishing in systematics about the question of quality control over information published on the Web. In the *Philosophical Transactions of the Royal Society* issue on "Taxonomy for the Twenty-first Century," Forey et al. (2004) felt that the absence of mechanisms for peer review meant that Web sites were not appropriate outlets for primary publication and should generally be used only for dissemination subsequent on peer-reviewed publication. Beyond that, they felt that the most useful and credible Web sites would gradually emerge by processes recognizable from other media: "The best taxonomy has always come from individuals with the experience and breadth of knowledge to provide an authoritative overview, and it is our belief that data-rich and scientifically useful sites will become self-evident to the wider community" (Forey et al. 2004: 651). Scoble (2004) too sees quality control as a major issue for taxonomy on the Web, both as far as individual sites are concerned and in the emergence of a usable infrastructure out of diverse sites. He, like Knapp et al. (2002), sees developments as evolutionary and emergent: "Like Web content generally, taxonomy on the Internet has arisen idiosyncratically, but the more taxonomy that is posted the sooner will an infrastructure emerge, simply by virtue of volume" (Scoble 2004: 702). Along these lines, Scoble finds it hopeful that various projects (described in chapter 6 as metainitiatives) are working toward consolidating different information sources or providing easier access to distributed databases. He also discerns a "mood," among at least some taxonomists, that an inevitable shift is happening from paper-based to Web publishing. Scoble's view of change in taxonomic communication, like many of the contributors to the "Taxonomy for the Twenty-first Century" special issue, is of a system that will change inevitably, but gradually and conserving much of the ethos of the previous version. The mood, in considerable contrast to the initial bold visions of Godfray (2002a,c), is one of continuity and emergence rather than radical change or top-down solution.

Although systematists see some clear potential advantages to the use of electronic communications in systematics, the formalized communication structure provided by nomenclatural codes builds considerable inertia into the system and make rapid response to new developments unlikely. This inertia is typically portrayed by systematists in largely positive terms, as allowing for a reasoned consideration of moves toward a communication system that retains the strengths of the existing system, and in particular retains the long-lived archival qualities that systematics values so much. Systematics is not yet ready to transform its practices in a radical way. The air of urgency that pervades systematics promotes the turn to electronic publishing, but is countered by the concern with permanence and the feared ephemerality of digital information, as well as uncertainty about quality control.

Overall, the extent to which systematics has incorporated virtual technologies into its communication culture depends on its ability to retain the qualities it values in the published literature such as the concern with permanence, combined with the will to exploit desirable qualities to extend or reinforce existing practices. Where these technologies are being used, it is often to build on or extend existing practices such as community-based information services or indexing facilities. Here, virtual technologies are embraced as a means of extending access and also enhancing value, often by giving the community increased opportunities to amend data. Radical opportunities either to incorporate explicit centralization, via registration of names or consensus classifications, or to provide wider access to publishing opportunities by allowing names to be validly published by electronic means, have not as yet been taken up. Current moves suggest that change will need to be viewed as incremental rather than radical, and that any drastic changes are likely to attract considerable opposition. Systematics has adopted an explicitly democratic approach to decisions on the formal aspects of its communication culture, and this makes radical change in the short term unlikely. Indeed, to be politically acceptable to this community, change needs to be portrayed as incremental and appropriately rooted in heritage.

Taxacom and the Disciplinary Constitution

One of the factors that has made systematics a fascinating and rich field of ethnographic enquiry is the tradition of reflexivity and consciousness of disciplinary identity that characterizes the discipline. Having a centralized approach to nomenclatural issues orients members of the discipline toward one another, and means that there is an awareness of the need to shift the whole discipline at once where nomenclatural matters are concerned. There is also a tradition of philosophical examination of methods and concepts that is, for many members, richly interwoven with the practical tasks of classification and naming. Systematics also has a history of incorporating new techniques and forms of data, and of organizing programmatic conferences and seminars to examine the need for change (Huxley's *New Systematics* [1940] serves as a good example, as do more recent attempts to examine the problem of nomenclatural instability such as Hawksworth 1991).

While conscious of common endeavor and oriented toward one another as a wider community, systematists are also isolated in their concentrations on particular groups of organisms, and focused to some extent on locally significant institutional matters. At the same time as it has developed a collective tradition, taxonomy has also traditionally been a solitary discipline. When I examined the publication landscape of the Collembola I also studied lists of practitioners drawn from conference attendance. Analyzing the resulting data on the distribution of Collembologists between institutions showed that practitioners were more dispersed than might have been expected on a random basis. Not only did they not cluster together in research groups, it seemed that employing one Collembologist made an institution less, rather than more, likely on average to employ another at the same time. As specialists within a field that provides far more problems than can be tackled by the personnel available, taxonomists have always tended to work alone. Exceptions might be the teams who work on projects such as large-scale floristic inventories, although even here work is often parcelled out among nonoverlapping specialists. Taxonomy is, then, both a collective endeavor with a disciplinary identity and a highly solitary activity with high levels of autonomy and little opportunity for frequent face-to-face interaction with experts in closely related areas.

Against this backdrop, a form of communication has emerged through which taxonomists regularly approach one another, at a distance, with requests for assistance and expertise. As well as the distributed taxonomic instrument which operates at the level of specimen collections, as we saw in chapter 4, there is a distributed network of taxonomic expertise, which previously operated via correspondence and face-to-face conferences as well as the formal journal literature. This existing reliance on communication at a distance, together with the reflexive orientation of practitioners toward the goals and status of the discipline, appears to make it ideally placed to exploit the possibilities of computer-mediated communication. In this section I explore the form that computer-mediated communication currently takes for systematists, and examine the extent to which the observable online activities are a reasonable representation of the discipline. I sought to find out how systematics made something distinctively its own from computer-mediated communication, and ended up focusing on how that in turn helped to remake the sense of what it was to be a systematist.

Systematists increasingly rely on computer-mediated communication: as Winston says in her introduction to species description, "the best way to communicate with a taxonomist today is usually by email" (Winston 1999: 38). There is a wide range of specialist mailing lists aimed at disseminating information on given groups of organisms, techniques, projects, or software. These operate with varying degrees of success, some flourishing and becoming an important part of working in that area, others failing to reach a critical mass of usage to become significant and often dying away. One list that appeared particularly successful and significant for the promotion of electronic resources in systematics was the Taxacom list. It is long-lived in Internet terms: members trace the origins back to a bulletin board set up by Richard Zander in 1987. Later, two lists, taxacoma and taxacomb, were based at Michigan State University. In its currently recognizable form it dates back to 1992, when the list was transferred from Michigan State University to Harvard—it is from this date that archives are currently available, and thus this period on which I base my account of the group, including a subsequent move to the University of Kansas Natural History Museum.

The list was originally focused on computer-related topics. The announcement of the list's move to Harvard specified its scope thus: "The emphasis of the new Taxacom@harvarda mailing list will continue to be on technical aspects of collections computerization, network access issues, and on software for systematics research" (Beach 1992). Nonetheless, over time the list has turned into a more broadly focused forum for discussions and announcements, which still contains many references to electronic resources but also encompasses debates about classificatory theory and technique, and topical issues in the discipline, and non-computational job announcements. It has become more a site for discussing taxonomy via computer than a list focusing on computational aspects of taxonomy.

I have tended to use the Taxacom list in the ethnographic parts of the current study as a window on the concerns of the discipline, allowing me to judge the mood of current debates and the level of acceptance of new proposals. As an informal yet public arena, Taxacom provided a good supplement to the more focused and formal occasions provided by face-to-face interviews, and gave what seemed a more "natural" feel for the concerns that practitioners face. It is, however, problematic to treat any online forum that one comes across as a straightforward reflection of the way that the world is. Online interactions do not straightforwardly reproduce offline concerns. As I outlined in chapter 2, many factors may intervene to shape online environments for science in particular ways, from the beliefs that members hold about how they should be used, to practical and economic constraints that exclude some potential participants altogether, and to the conventions that members collectively develop when those of them who can come together online do so. It is therefore important from a methodological perspective to assess just how far the Taxacom list could be said to represent the discipline of systematics and what biases shape its divergences from the other ways in which the discipline is enacted. This also forms a key point in the overall argument of this chapter. If disciplines are constituted through their cultures of communication, it is important to ask where the new computer-mediated forms of the discipline are continuous with its past, and where they represent new opportunities or pitfalls.

My analysis of the Taxacom group is based on a range of sources of data. The archive is publicly available (http://listserv.nhm.ku.edu/archives/taxacom.html/) in searchable format, and archive files are also available for download. I explored some basic patterns and statistics of the group by parsing the archive files into a simple database. I looked at the profiles of prolific and less active senders of messages to the group, and explored the temporal patterning of interactions. I was also provided by James H. Beach, the list coordinator, with analyses of the membership lists, including a breakdown by country and a list giving the date of joining for a number of members. I have subscribed to the list since the beginning of the study, and therefore also have a more real-time ethnographic sense of the pace of activities, the general tone of interactions and the extent to which other events in systematics are reflected within the group. Where I quote directly from messages to the list I have asked permission from the authors: this seemed appropriate since these authors could never have expected their words to be used in this context, and I had no wish for them to feel that their current professional reputations were threatened by past words coming back to haunt them. Asking permission was also often a useful engagement, leading to extended correspondence on the use of the list or on particular substantive topics.

In addition to my own pattern searching and experiential assessments of the group, it was vital to find out what those concerned made of it all. I therefore asked the coordinator's permission to send a message to the whole group, which was duly received. I sent messages asking for members' assistance to tell me how useful they found the list, how they used it, what they felt about its reflection of the discipline's concerns, and what other lists they belonged to. The full text of the messages I sent is given in the appendix. I received twenty-five replies, in varying lengths, many giving me detailed accounts of thoughts about the list, its importance and biases. Each message was acknowledged with thanks and questions for clarification, and some led to a more extended correspondence. I was delighted to receive this number of responses, which by a long way exceeded my previous experience of asking for input from newsgroups. I use quotes from these comments to illustrate my description of the list, leaving these quotes unattributed to preserve the confidentiality of the authors, a number of whom explicitly asked that their comments be treated as "off the record." Nonetheless, the list had some 1,400 subscribers and my response rate of nearly 1.8 percent gave me little grounds for complacency in my ability to make unequivocal statements about what the group meant. The observations I have to make about the Taxacom group cannot therefore be said to represent to any extent the whole truth about its nature. I capture some salient aspects to consider what a forum like Taxacom could mean for the discipline as a practical outlet and a component of its sense of self, leaving open the possibility of diverse other ways of working with the group that remained invisible to me.

In recent years the list has, as I remarked earlier, become more general than implied by its original focus on computerization, network issues, and software. Common topics for messages include the promotion of new electronic resources, but also extend into much broader areas. Job advertisements are common; many do not have a particular focus on computational skills but are rather focused on collections management or systematic projects without explicitly topicalizing information technology. The list has also served a valuable role for many as a source of practical help, including advice on software but extending into searches for literature, for specimens needed in connection with a particular project, for legal advice, and for information on the whereabouts and contact details of colleagues. Appeals for help and support, where a particular collection is thought to be under threat, or a systematist feels wronged by the behavior of an institution, have also been made via the list. Finally, and significantly for many of the people who gave me their opinions about the list, it has become a forum for debates about concepts including the techniques and foundations of phylogenetic analysis and the role of molecular data. Much of the list content is made up of stand-alone announcements, or consists of short questions and answers. Debates, however, often extend into much longer threads of statement and counterstatement.

The remit of the list seems to be fairly organic, by which I mean that it has been allowed to become what its members want it to be. Suggestions that topics are inappropriate, or that discussions should be taken off-list, have been fairly rare recently, giving the public perception at least that the list as a whole is comfortable with this emergent broad remit. Exceptions are where a discussion is felt to have gone on too long, or to have developed an inappropriately personal tone. There have been past discussions on whether the list should be moderated, but these have focused more on the need to guard against spam and "unsubscribe me" messages accidentally sent to the list rather than concerns about systematics-based off-topic messages. A vote on moderation in 1995 produced a majority in favor, but the coordinator opted instead to introduce various technical measures (such as rejecting messages with no subject line, not allowing non-subscribers to post messages), and the list is set by default to direct replies to the individual sender rather than to the list. Currently the low levels of complaint suggest that for most members this solution is working.

Despite the large list membership, debates tend to be populated by a much smaller cast of "regulars." A far wider cast use the list to make announcements, while the majority hardly ever, or never, post messages. There is some list turnover (and occasional fears expressed that excess spam causes people to unsubscribe), but there are many long-term members. The statistics of list usage are depicted in figure 5.1, showing the numbers of messages posted and of unique email addresses sending messages during a ten-year period. All of the figures that I present here are subject to a certain amount of latitude, partly thanks to the vagaries of the approach I used to parsing which may have occasionally failed correctly to recognize a message. More significantly, the count of the number of different people who sent messages to the list each year is inflated since one individual might have used several variants on email address format, and each would have been counted separately. Nonetheless, the figures are good enough to give a sense of the liveliness of the list, averaging around six messages a day since

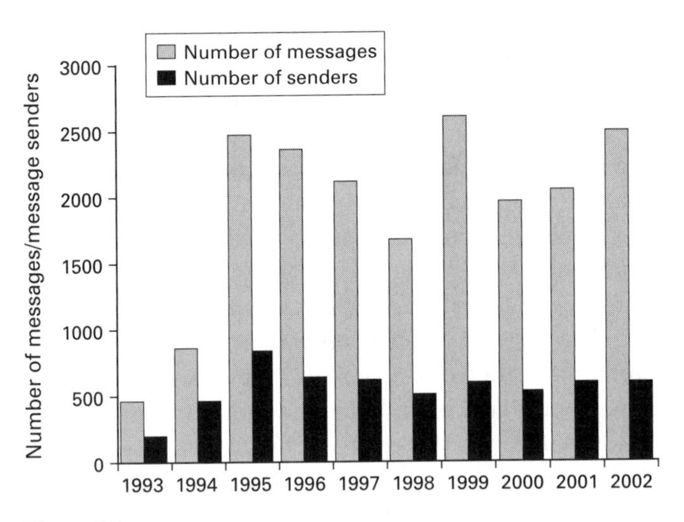

Figure 5.1
Bar chart showing numbers of messages sent to Taxacom, and the number of unique addresses sending messages each year 1993–2002.

1995 and drawing in around six hundred different contributors each year. The figures suggest that this is a mature and relatively stable list, an impression supported by the people who responded to my queries who found subscribing generally to be a useful and rewarding experience. Of course, there is no absolute criterion of a "healthy" list, but that people keep coming back, forming a strong cohort of long-term, satisfied users, would seem to be a fairly reliable basis for making that judgment.

The list, then, is relatively unpoliced as far as topic is concerned, and since anyone is allowed to subscribe without being vetted in advance it is also open in terms of membership. To that extent it reflects the discipline. This, though, is only a part of the picture. Several of the people who responded to my inquiries about the list mentioned a North American bias. According to the way that they perceived the overall international distribution of taxonomic activity, this list seemed overly focused either on people from North America, or on the concerns prevailing there. Geographical issues are, in contemporary systematics, highly politicized, as I noted in chapter 3. Although the Internet seems to have the scope to overcome geographic disadvantages by making information available to anyone anywhere, it is quite possible that lack of infrastructure and of skills will mean that the Internet instead exacerbates or reproduces existing divides. An emerging body of research suggests that the Internet, rather than creating a boundary-free scientific sphere, is tending to reproduce existing structures and divisions (Barjak 2006; Caldas 2006). There are, however, some indications that the Internet can help some scientists, particularly women in developing countries, to circumvent local constraints (Palackal et al. 2006). It was therefore important to make some effort to explore the geographic basis of the list to see how far it reproduced or helped to overcome the existing geographic structuring of taxonomic work.

Table 5.1 represents a breakdown of the list membership by country. In order to make some assessment of whether the membership of the list matched with the global distribution of taxonomists, I used data from the World Taxonomist Database maintained by ETI Bioinformatics (http://www.eti.uva.nl/tools/wtd.php/). Of course, this list itself is not independent of bias: since taxonomists are invited to sign up via the Internet, and much of the publicizing of the resource has been done via mailing lists, there will be some bias toward inclusion of taxonomists who are Internet aware. There have been considerable efforts, however, to make this list as comprehensive and geographically open as it can be, and it represents the best resource there is for assessing the global distribution of taxonomists. Membership of the Taxacom list can then be assessed as a ratio of the

Table 5.1
Breakdown of subscribers to Taxacom by country, ordered by percentage of the total taxonomists for that country and juxtaposed with GDP for comparison.

Country	Number of subscribers to Taxacom[i]	Number of taxonomists registered on ETI directory[ii]	Percentage of ETI registered taxonomists subscribed to Taxacom	GDP at Purchasing Power Parity per capita[iii]
Hong Kong	4	3	133	32,292
Nicaragua	1	1	100	2,779
Slovakia	6	7	86	16,110
USA	707	917	77	41,557
Iceland	2	3	67	35,686
Australia	105	191	55	31,020
Japan	17	31	55	31,384
Dominican Republic	1	2	50	7,055
Namibia	1	2	50	6,658
Yugoslavia	3	6	50	*
Canada	60	131	46	34,444
Estonia	7	16	44	16,461
Great Britain	82	198	41	30,309
Finland	10	25	40	30,818
Singapore	2	5	40	28,228
Switzerland	19	48	40	33,168
Denmark	11	28	39	34,718
New Zealand	18	53	34	24,805
Austria	15	45	33	32,962
Slovenia	3	10	30	21,695
Sweden	18	60	30	29,537
Chile	4	14	29	11,537
Belgium	24	86	28	31,549
Germany	59	212	28	30,150
France	32	134	24	29,203
Italy	24	101	24	29,414
Netherlands	28	118	24	30,363
Costa Rica	4	18	22	10,316
Croatia/Hrvatska	1	5	20	12,364
Ireland	2	10	20	40,003
Lithaunia	2	10	20	14,198
Hungary	3	16	19	16,627
Czech Republic	6	36	17	19,475

Table 5.1 (continued)

Country	Number of subscribers to Taxacom[i]	Number of taxonomists registered on ETI directory[ii]	Percentage of ETI registered taxonomists subscribed to Taxacom	GDP at Purchasing Power Parity per capita[iii]
South Africa	12	75	16	11,035
Brazil	25	172	15	8,745
Indonesia	1	7	14	3,853
Russian Federation	12	84	14	11,209
Ecuador	1	8	13	4,010
Latvia	1	8	13	12,886
Mexico	17	126	13	10,090
Korea	2	19	11	22,543
Turkey	5	52	10	7,958
Venezuela	3	32	9	5,801
Iran	1	13	8	8,065
Norway	2	26	8	41,941
Philippines	1	12	8	4,770
Spain	10	140	7	24,803
Thailand	1	14	7	8,542
Ukraine	1	15	7	7,182
Colombia	2	31	6	7,303
Poland	2	33	6	13,275
Argentina	3	74	4	13,153
China	2	56	4	6,193
India	4	134	3	3,262
Portugal	1	34	3	19,949
Taiwan	1	109	1	27,122
Algeria	0	3	0	7,095
Armenia	0	1	0	4,048
Bangladesh	0	6	0	1,998
Belarus	0	1	0	8,186
Belize	0	1	0	7,635
Benin	0	2	0	1,147
Bermuda	0	2	0	*
Bolivia	0	6	0	3,049
Brunei Darussalam	0	1	0	24,826
Bulgaria	0	9	0	9,205
Cameroon	0	1	0	2,284

Table 5.1 (continued)

Country	Number of subscribers to Taxacom[i]	Number of taxonomists registered on ETI directory[ii]	Percentage of ETI registered taxonomists subscribed to Taxacom	GDP at Purchasing Power Parity per capita[iii]
Cuba	0	15	0	*
Egypt	0	11	0	4,282
Fiji	0	1	0	6,282
French Guiana	0	1	0	*
French Polynesia	0	1	0	*
Gabon	0	1	0	6,977
Georgia	0	1	0	3,038
Greece	0	12	0	21,529
Greenland	0	1	0	*
Guatemala	0	8	0	4,136
Guyana	0	2	0	4,685
Honduras	0	1	0	2,793
Israel	0	10	0	22,944
Jamaica	0	2	0	4,471
Kenya	0	7	0	1,125
Kyrgyzstan	0	1	0	2,061
Laos	0	4	0	2,049
Lebanon	0	1	0	6,205
Libya	0	1	0	11,354
Luxembourg	0	3	0	66,821
Malaysia	0	24	0	11,160
Malta	0	2	0	20,015
Mayotte	0	1	0	*
Morocco	0	2	0	4,444
Nepal	0	5	0	1,471
Nigeria	0	5	0	1,188
Oman	0	1	0	16,300
Pakistan	0	11	0	2,567
Panama	0	4	0	7,327
Papua New Guinea	0	1	0	2,414
Paraguay	0	3	0	4,663
Peru	0	12	0	5,594
Puerto Rico	0	8	0	*
Romania	0	12	0	8,258
Sri Lanka	0	9	0	4,145

Table 5.1 (continued)

Country	Number of subscribers to Taxacom[i]	Number of taxonomists registered on ETI directory[ii]	Percentage of ETI registered taxonomists subscribed to Taxacom	GDP at Purchasing Power Parity per capita[iii]
Sudan	0	1	0	2,417
Syria	0	3	0	3,871
Trinidad and Tobago	0	1	0	13,958
Tunisia	0	3	0	8,223
Turkmenistan	0	1	0	7,854
Uruguay	0	12	0	9,619
Uzbekistan	0	1	0	1,834
Vanuatu	0	1	0	3,397
Viet Nam	0	7	0	2,782
Yemen	0	1	0	745
Zimbabwe	0	2	0	2,413
Kuwait	1	0	*	16,297
Malawi	1	0	*	596
Uganda	1	0	*	1,817
Zambia	1	0	*	911

[i] Data provided by James H. Beach on 21 May, 2004 using "Review by country" listserv command.
[ii] Data taken from the ETI World Taxonomist Database (http://www.eti.uva.nl/tools/wtd.php) on 12 August 2005).
[iii] Data taken from International Monetary Fund World Economic Outlook Database for 2005 (http://www.imf.org/external/pubs/ft/weo/2005/01/data/index.htm).
* Data not available.

overall number of taxonomists in a given country given by the ETI database. In the final column of the table I give the adjusted GDP for each country, to give some basis for assessing whether economic advantage and level of development shape the differences in list membership.

The table is sorted so that those countries with the highest ratio of their taxonomists subscribed to the list appear at the top. The United States does indeed appear high on the list, with some 77 percent of the taxonomic population involved, and in absolute terms representing approximately half of the overall number of subscribers. It is thus not surprising that list users should feel that it was dominated by the US. The table also reveals a broad relationship between the percentage of taxonomic population subscribed to Taxacom and the adjusted GDP (with a Pearson product-moment correlation coefficient of 0.47), suggesting that differences in economic advantage do indeed leave some groups of scientists less likely to use the list. There are, however, some surprises and of course some quirks among the figures. Among the countries with larger numbers of taxonomists but a lower GDP, Brazil appears higher than might be expected. Brazil is a highly significant country in terms of biodiversity, with good connections with systematics institutions in richer countries and a strong program of its own. Brazil is possibly more oriented toward the Internet and specifically the Taxacom list than might be predicted on economic indicators alone. Taiwan, on the other hand, appears low down, excluded from participation possibly for linguistic reasons, or from a lower level of preexisting links to other countries.

The taxonomists on the ETI database are largely self-identified, as are the members of the Taxacom list. It is possible, then, that we are looking at different populations. Certainly this is true of Hong Kong, where more are subscribed to Taxacom than appear in the ETI database. Several people who emailed me to give their thoughts on the Taxacom list identified themselves as outsiders or ex-taxonomists who used it to keep up with what was going on in the field, and therefore would presumably not have registered themselves in the ETI database. The figures do not, therefore, merit too detailed an analysis. Nonetheless, from this crude data set we do have some grounds to suspect that the Taxacom list is in practice not equally available to all taxonomists. Prior networks of connection, availability of network infrastructure, equipment and skills, cultural differences, and lack of linguistic common ground mean that the list is biased toward countries that are either in the more developed world or have strong connections with it.

The extent to which this level of geographic distortion matters is, of course, proportional to how important the list is in the overall disciplinary context. The list being what its members make it, there is always the potential that the list could be something quite different with an altered membership. There is certainly a missed opportunity for more direct engagement between the concerns of taxonomists in different areas of the world, and for those concerns to gain a more public airing than might otherwise be the case. There are also missed opportunities for taxonomists in developing countries to take advantage of the kind of practical appeals for help that the list enables. As it stands, as one member told me:

> There's a bunch of very vocal contributors who skew the demographics towards their interests [. . .]. But overall (if you remove their posts from the sample) I think the interests of taxonomists in the 'western world' are reflected by the discussions on taxacom.

The list is, then, largely seen to reflect systematics, not in an idealized version freed from inequalities of economics and geography, but in a form continuous with other portrayals of the discipline. A number of respondents to my queries reflected this kind of continuity in their assessments of what Taxacom meant to the discipline, suggesting that systematics would continue relatively unchanged in the absence of the list. To that extent, those not participating are seen as missing out on a useful practical tool rather than being excluded from disciplinary participation altogether, although it is, of course, problematic to make that assumption on their behalf.

I have tended to use the Taxacom list as a mirror of the discipline. It became clear from the answers that I received to my queries about the list that other subscribers were using it in this way too. Several people were ex-taxonomists, or graduate students, who portrayed the list as a way of looking in on the discipline, checking up on what the current mood was, or learning about issues by watching debates. Debate was seen by some as productive, in itself, by promoting a questioning attitude of mind:

> It's very easy to live in your little world and take a lot of things for granted or unquestioningly accept that "this is the way it's done." Taxacom by definition works beyond those sorts of boundaries and I think it is a very powerful tool for exposure, especially of students like myself, to different approaches. I also think lively debate keeps a discipline alive.

Institutional perspectives were portrayed as inevitably limited, by contrast with the "broader perspective" offered by Taxacom, or the access to a "world of taxonomy outside my department." Taxacom served, for several, as a kind of current awareness service, "a good way to absorb a little of the 'culture' of taxonomists," and as an easier or more transparent route than the formal literature:

> I deal with taxonomists, curators, database managers and users of taxonomy on a day-to-day basis but rarely get a chance to sit in the library to read the new literature. Listservers, including Taxacom, partially fill the void in that they alert me to some of the topics that are exercising the minds of the people I work with or to whom I provide funds.

The list was also contrasted favorably with other means of informal communication:

> It's a good way to network and find out how people are dealing with problems. This is so much easier than the phone or the rumor mill.

For one respondent it was seen as a hybrid form:

Both scientifically "respectable" (it appears to attract a substantial number of professionals) and outside the constraints normally imposed upon scientific discourse through the institutions of science (particularly publications).

The style of Taxacom posting was therefore approved by many as making it useful for learning about particular issues by access to different sides of a debate, and by a certain directness of style that meant issues of importance were addressed in a scientifically respectable form more succinctly, and more directly, than the formal literature allowed.

The list was, therefore, treated as a way of keeping up with the discipline. This is not to say, however, that readers were naive in their assessments of messages that they read, or took the outcomes of debates at face value. A variety of sources of bias were noted by respondents, most often that the list tended to be dominated by the concerns of a few more vocal individuals. Other biases noted included the lack of representation of modern taxonomic practices, particularly molecular approaches, and a broader tendency to focus on issues that were controversial:

Generally I found that more marginal issues and minority views tend to get excessive space, whereas more general issues and widely accepted views tend to get missed out.

This kind of skepticism about the biases of the list did not prevent others from using it to situate themselves within the discipline. One reader wrote of using the list to work out how far his own views were shared, and of gaining a different view to that from the literature:

I find it very interesting—especially to find that, in certain controversial areas, I am not alone. Indeed, the number who hold certain views seems much greater than one would expect from the published literature.

In general, responses suggested that readers combined the two approaches, reading the list for a perspective on developments in the discipline but tempering that reading with their perceptions from other sources, including the journal literature, interactions with colleagues, and their own specialisms and experience. Reading the list was an occasion for learning about taxonomy, but not to be taken in isolation.

By reading the list as, at least in part, a reflection of the discipline, users also shaped their practices in terms of posting messages. Several people who replied to my queries mentioned a reluctance to post, based on a recognition of the high stakes involved:

My feeling is it's the sort of list that you don't want to post too frivolously to on if you want to be taken seriously (some people do, then don't seem to come back . . .).

One spoke of liking "the anonymity of silence," while another attributed reluctance to post to wishing to avoid "making a fool of myself in front of everyone." The sense that the list was a disciplinary forum made for inhibition, since it was felt that posting to the list would affect one's disciplinary reputation. For similar reasons several respondents asked me to keep the source of their comments confidential, or not quote from them directly, since they felt that their remarks would either offend the discipline or list, or affect their future chances of employment.

If this sense of being public in front of the discipline inhibited some from posting, for others it acted as an incentive. One respondent described this as an irresistible urge to contribute on a topic that one knew about:

> I definitely think of myself as a "lurker" on these lists—posting very rarely and only in response to posts that I can't resist not responding to.

Another put their motivation for posting more succinctly as "fighting obscurantism ;-) !!!" Judgments about whether to reply to the list or privately to the individual were described by another as made on the basis of careful assessments, again founded on an appreciation of the list as a public forum to be approached with respect:

> I do this only occasionally, such as this response! If so it is either because requested to do so, as in your case, or because I think my response would be of much more interest to the person it is directed to than the general list, or because it may be something I know little about and it seems better to ask for clarification from a known expert than cluttering up the list, or (rarely) because I think someone has posted an erroneous statement and would probably appreciate being corrected privately more than in a public posting.

These forms of reasoning compare with the results found by Lakhani and Hippel, when they studied the provision of "mundane but necessary" online help by an open source software community (Lakhani and von Hippel 2003). They suggested that the provision of answers was very little overhead to those who were already reading the forum in order to update and expand their own knowledge. Reasons given for providing answers to queries ranged from the wish to serve a general sense of reciprocity, or the open source movement in particular, to a concern with reputation and being seen to be knowledgeable or helpful. It seems likely that the rationales of people who provide information in response to queries on Taxacom might not differ dramatically.

The impression from reading messages to the Taxacom list is of a taxonomic community, as a viable place for generalised reciprocity, as well as a place where one could build reputation. This is not, however, likely to be interpreted in the same way for all list members, and for some participation may not be a priority at all. Matzat (1998) looks specifically at scientific usage of discussion groups for informal advice and develops a differentiated model for the likelihood of asking and answering questions, which depends on factors such as the status of the scientist concerned, the other forms of advice available to them, and their other time pressures. Hard-pressed and high-status scientists may find themselves too busy to participate, able to access plenty of other sources of advice, and aware that reputation could be reduced rather than enhanced by online exposure. It is important, then, not to take the idea of reciprocity as determining individual responses to online participation. These rather earnest pictures of list activity as focused on evaluations of potential gains should also not be overstated: it was clear that participants in some of the more intense debates found it to be an enjoyable experience, and one framed posting messages as an opportunity to "stir the pot a little."

In stylistic terms, participation on the list was often portrayed as continuous with other spheres of communication. As one reluctant poster described it:

> I respond to postings only rarely, and prefer to take up issues with individuals rather than via the Listserver itself. In this I don't think that I behave any differently to the way I dealt with things in the days of snail mail, conferences etc. I was always more likely to collar someone after a meeting than challenge something they said on the floor.

In gender terms, one respondent suggested to me that women were underrepresented among posters to the list, and I followed up by asking the list as a whole whether this were the case and if so why. Responses tended again to tie the list into other spheres of scientific communication, suggesting that it was "the whole Alpha Male Thing," or that science operated according to "boys' rules," or that the visible participation in Taxacom was the upshot of a set of factors:

> Given that women have less hormonal drive for confrontation; are fewer in science, systematics and on Taxacom than are men; and that they have less time to spend on written conversations in an overall sense, it is not at all surprising that contributions on Taxacom from women are many fewer than are those from men!

I opted not to conduct a detailed analysis to attempt to determine whether women were indeed underrepresented on the list as compared to the discipline as a whole. Besides the considerable problems of acquiring adequate data, it seemed that the issue was sensitive to many of those concerned, and it seemed inappropriate to intervene by providing an analysis that purported to confirm or deny perceptions about list bias in this regard. In this I have been inconsistent with the decision that I made to attempt an analysis of the geographic biases of the list. In that case at least better data were available, but gender offered a minefield in which there was little justification to intervene with imperfect data. The key point to take away is the sense of participants that the list was continuous not just with the substantive concerns of the discipline but also with its working conditions and stylistic preferences.

The more prolonged interactions on the Taxacom list could be seen as continuations of the existing tradition in systematics, and science more broadly of a combative style of debate, which can be productive in encouraging clarification and extension of principles. Hull (1988) explores the dynamics of competing approaches to systematics and highlights the crucial role that conferences and journals play in providing occasions for positions to be specified and brought into juxtaposition with one another and for allegiances to form, although he possibly overemphasizes the adversarial nature of debate. A similar adversarial style is noted by some readers of the Taxacom list, in their descriptions of debates as learning experiences. Discussion both unifies and fragments the discipline/community. It provides a public space for discussion, so that people can read and respond to it as "what the community is concerned about." But it also provides a very dynamic space in which differences can be identified and played out, carving up "the community" into a fractured and sometimes oppositional territory. I did not see any evidence of concern about issues of credit and priority in discussions on Taxacom (although as Hull 1988 says, grumbling about priority is rarely done publicly). Contributions tended to be positioned as opinions and clarifications, or to be linked to other sources such as published papers or Web sites. The treatment of credit and priority assumes that the list is tied into the other communication venues of the discipline, suggesting that appropriate credit will accrue in those other venues if information offered on the list turns out to be useful.

Taxacom does not stand as an isolated space. It is tied in to disciplinary practices in a range of ways. Most obviously it is tied into the working practices of individuals by their weaving it into daily routines and their communication with co-present colleagues. People spoke to me of keeping an eye on Taxacom, making skimming messages for interest part of their daily routine. They knew that others did so too, although they did not necessarily know who, exactly, did subscribe, thanks to the large scale of lurking:

> I have noticed that lots of colleagues do subscribe but never send anything, because whenever I have sent something they reveal themselves by saying "I saw your note on Taxacom"!

Taxacom thus became part of informal networks of advice and gossip, as it was both the subject of conversations and the material through which ties were reinforced:

I can think of one "old-school" colleague who wouldn't dream of subscribing to an email discussion list (although very_au fait_with email), but is always very interested when I pick up snippets and pass them on.

Also, people I interviewed who did not read the list themselves spoke of having messages thought to be of interest forwarded or mentioned to them by colleagues.

Messages to Taxacom also point systematists toward other resources and spaces. As I have described, a common use of the list is to advertise a new version of an electronic resource, inviting readers to visit and check out the new offering. Other forms of interactive space have also been tried out within systematics, although with limited acceptance so far. Many computer scientists are involved with the standards development efforts of contemporary systematics, and have brought in communication conventions from that community. For example, as a part of their standards development activities, a team at Napier University developed a wiki[3] for presentation of the taxonomic data schema they were developing and invited users of Taxacom to visit it to make their comments on the schema. The resulting space was used, it appears, only by a few individuals most closely concerned with the process, much of the work of developing and negotiating the standard going on through more established routes like mailing lists and emailed responses, consultative visits, conferences, and workshops. The mailing list for the comments on the emerging standard shows a considerably wider cast of contributors than the wiki. The wiki, then, had limited success as a presentation of the full spectrum of reactions to the schema. A wiki might be often portrayed as an easy means of developing inclusive debate, but clearly outreach is still needed to stimulate interest and commentary and many potential contributors still prefer other means to make their point.

Despite the limited success of the wiki, to date, as a way of collective working for systematists, the invitation to Taxacom readers to visit the wiki space shows two facets of the emerging communication landscape of systematics. On the one hand space is highly differentiated, and there is a strong sense of the purpose of each space. While Taxacom is a way of addressing the systematics community in a broad sense, a diverse array of other spaces is emerging, constituted through special-purpose mailing lists, wikis, and Web sites. On the other hand, spaces are richly connected through their mutual visibility, systematists being invited into different spaces and becoming aware of them through their standard awareness practices of monitoring key sites and communicating with colleagues in related areas. Taxacom is a valuable site for the constitution of a disciplinary identity and has been constructed as a disciplinary space, but it forms only a minor part of the communicative resources needed for an effective career in systematics. The special-purpose sites, the formal publishing system, the institutional communities, and specialist networks and conferences together form the basis of a rich and varied communicative landscape whose connections a successful systematist must navigate.

Taxacom gives, to some extent, the systematics community a sense of what it is. Its importance for the analysis presented in this book is twofold. In the first sense, this form of computer-mediated communication offers up a renewed disciplinary reflexivity, building on the existing tradition but extending it in terms of immediacy of feedback and broad public accessibility. In a second sense, it has assisted in subtly shifting the discipline's sense of what it is in substantive terms. The use of the list for announcements of new electronic resources has helped to reinforce the role of these resources as a central part of the discipline's efforts, and to normalize production of Web-based resources as a core part of disciplinary endeavor. Inasmuch as readers take the list to be a reflection of the concerns of the discipline and the announcement of new resources is a routine activity on the list, it has clearly become everyday for systematics to concern itself with the production of electronic resources. Taxacom has not driven this activity, but it has provided a venue for its practitioners to come together and to promote their products to the wider discipline.

Conclusion: Communication in Systematics as a Special Case

This chapter has shown that the communication system, and the potential for transformation in that system, is shaped in systematics by references to the prior history of the discipline and in particular its very formalized approach to the valid publication of new names. At the same time, the geographic distribution of systematics and a strong reflexive orientation toward the concerns and status of the discipline predisposes systematists toward the use of informal computer-mediated communications. These informal communications have provided a place for solidifying ideas about the nature of the discipline, and have also acted as a site for normalization of electronic resources within the practices of the discipline. The potential of the mailing list as a site for enactment of disciplinarity is enhanced by understandings of the nature of systematics as a relatively homogeneous discipline in terms of method and theory: while there are special issues related to each group of organisms, ideas about evolutionary descent cut across the whole discipline and are seen to provide the basis for systematists across specialisms to hold meaningful conversations. There is thus more basis within systematics for a transdisciplinary mailing list than there would be in a diverse discipline like sociology where there is a consciousness that basic theoretical and methodological approaches can differ dramatically. In turn, of course, the Taxacom venue reinforces the sense of the discipline as having common concerns and holds it together to be experienced as a coherent entity in the face of other specialist mailing lists that might offer a more fragmented perspective.

Systematics is, then, quite a special case, and we should not expect the forms of communication that systematics develops necessarily to transfer to other fields of research. As other commentators have found, there is a disciplinary specificity to new forms of communication in science (Nentwich 2003; Fry 2006; Merz 2006), and models may not translate between disciplines. Aspects of the prior practices of the discipline, its expectations about coordination, and its ability to innovate will shape its willingness to adopt new models and the form those models take. For these reasons there is a strong contrast between the forms of communication I have described for systematics and the models seen in other areas of e-science. One aspect of systematics that shapes the very possibility of new modes of communication in systematics is its current political orientation toward open communication with a broad and relatively undefined user base. While many e-science areas are working on a collaboratory model, involving relatively closed, secure, and password-protected areas for sharing data, instruments, and communications, systematics has developed a more diffuse and open system of databases. It would be of limited political viability and funding appeal for systematists to develop closed access systems.

In sum, communication regimes have disciplinary specificity and at the same time are significant conduits for the enactment of disciplines. The discipline both constructs and is constructed through its communication practices. Within systematics, new communications technologies have been embraced as ways of renewing the discipline, but in forms carefully shaped to be seen to retain valued qualities of prior media. The new communications technologies have provided a means of imagining the past and future of the discipline, and for exploring the relationships which past and future should have with one another. In this model new media do not simply replace the old. The deployment of computer-mediated communication sits alongside other media, each of which has a place within the practices of the discipline. Without ever appearing to constitute a radical transformative force, computer-based media have suffused the discipline and allowed it to reinforce its sense of self.

In chapter 3, we saw hopes for the transformation of the discipline on symbolic and practical levels through the deployment of new ICT-enabled communication regimes. These visions tend, it seems, to underplay the need for solutions to be experienced as making sense within disciplinary contexts and as a conduit for valued disciplinary practices. The need to be modern is balanced out by serious concerns with the role of heritage. A further aspect of new communication regimes that needs to be taken into account in assessing their viability is the source of the labor that will make them happen. Mailing lists like the Taxacom list have been successful partly thanks to their incremental, grassroots-style organization. Although the list is hosted by an institution this is not experienced as a major burden beyond the care required from a list owner in managing subscriptions and providing a contact point for problems. Individuals sign up to the list and make it live through their contributions. Major resources like nomenclatural lists or specimen collection databases require a lot more organized labor to make them happen in credible fashion, as we saw in chapter 4. Beyond disciplinary acceptability, then, we have also to think about issues of institutional and individual labor investment, and the returns that might accrue from instituting such an innovation in the communication practices. Chapter 6 turns to this issue, mapping the field of systematic cyberscience in terms of the entities that populate it and the ways in which their various investments become meaningful to make. The discipline is populated and enacted by institutions, initiatives, and individuals, and each of these has a distinctive part to play in structuring the discipline as a cyberscience.

6 Individuals, Institutions, Initiatives

When I surveyed the scene of systematics in chapter 3 to find out who was most closely involved with developments in ICTs and what their concerns were, I used the House of Lords Select Committee report from 2002 as a proxy to map the field for me. I took it as self-evident that those who volunteered evidence to that committee were identifying themselves as part of the field of UK systematics, and that those who were summoned to give evidence in person and quoted at length in the resulting report were being accepted as prominent commentators on the discipline. In this chapter I return again to the question of who is involved in the development of ICTs in contemporary systematics, looking at the issue in somewhat more depth than the House of Lords Select Committee report permitted, and taking into account the different pictures of the territory that alternative sources of data provide. In chapter 4 we saw that the role of the major institutions as repositories of specimen collections continued in the virtual age, augmented by digital resources but not, according to most of those concerned, threatened in their status as the absolute recourse of a systematics concerned to record and understand the variation of living organisms. There are, however, new kinds of initiative coming into being, propelled in part by the increasing significance of biodiversity on the global political stage, and digital resources offer the possibility to transform the institutional landscape built around material collections. In this chapter I will explore the landscape mapped out by traditional institutions and new initiatives, and investigate the extent to which they complement or compete with one another. I will also consider what roles this emerging landscape leaves for individual systematists to adopt, and how far work with ICTs has become an important component of the identity of a contemporary systematist.

As a complement to my own mapping of the territory of contemporary systematics I will also be looking at the ways in which this landscape is experienced and brought into being through the process of developing a new database. The example I use is an initiative that aimed to provide access to authoritative information about a particular group of plants. The project was developed as a means of providing a user-friendly interface for systematics, allowing nonsystematists to access information without needing to learn the details of nomenclatural complexities. For the initiative to be successful it needed to enroll both individuals and institutions, thereby enabling sharing of data and the collaboration of the key parties in a joint endeavor, and it also needed to identify a credible set of users for the information made available. I present a brief account of this project to explore the role of initiatives in defining the landscape of systematics, and to consider the role of branding in project promotion, and the dynamic tensions between the roles of institutions, initiatives, and individuals. In comparison with the fairly tidy version of the policy sphere presented by chapter 3, in which policy makers consider the evidence of institutions and make recommendations on which institutions act, this section presents a much more dynamic picture of the relations between initiatives and the policy sphere, mediated by concerns for fundability and sustainability, oriented to the actions of other initiatives and institutions, and populated by individuals with their own concerns and career goals.

In looking at progress of this one initiative it becomes clear that the success of projects can be greatly influenced by key individuals who advocate and support novel initiatives. At the same time, for many other individuals the project may only be a minor part of their role: they have to be persuaded to contribute time and data to an endeavor that is not a central part of their vocation. Both kinds of involvement can be crucial if a project is to be a success. Individuals also work for institutions, and this frames their possible response to initiatives. Whereas institutions have to sign up formally for initiatives, individuals within those institutions have to be found for whom participating is a meaningful thing to do, and indeed, often the direction of influence can reverse to involve individuals approaching their institutions for permission rather than institutions instructing individuals to participate. The tension between the careers of individuals and the interests of institutions, and the varying extent to which initiatives in ICTs are part of the conventional career path for individuals in systematics is explored in the final section of this chapter. I look at a variety of data sources, including published taxonomic work and project Web sites, to consider the extent to which work with ICTs is part of the daily environment of systematists as these sources represent it.

I begin the chapter with institutions, move next to initiatives, and end with individuals, but the key concerns along the way are with the tensions and dynamics between these different kinds of entity and the varying ways in which they experience the cyberscience of systematics. Combining these levels and foci of analysis allows me to revisit the question of branding, as raised in chapter 3, in light of the public face of digital resources, and see how credit accrues to individuals, institutions, or initiatives. Historically, reputation in science has largely been thought of as property of the individual scientist. It has even been argued that building reputation and hence accruing resources through a cycle of credibility represent the core activities in which scientists engage (Latour and Woolgar 1986). There is variation in how this cycle is organized between epistemic cultures, with molecular biology laboratories typically depending on a strong laboratory leader while high-energy physics operates through large collectives that design and report on experiments (Knorr-Cetina 1999). In the case of high-energy physics, Knorr Cetina argues that the experiment rather than the individual scientist is the brand, acting as a "collective epistemic subject" (Knorr-Cetina 1999: 168). The development of large-scale digital resources in systematics inevitably involves the work of many individuals. What, then, is the epistemic subject, or the brand, where online digital resources are concerned? Will this represent a challenge to traditional credibility and reward structures in systematics?

The question, as I have framed it above, revolves partly around whether there is anything special in the branding of digital resources, either simply because they entail effort from lots of people, or by virtue of their digital nature. On the latter point, Poster (2004) argues that, although the brand is supposed to remind us which company the product comes from, in practice when seeing cultural products consumers think more about the creative input of the associated actors, musicians, and authors than about corporations. In many ways this mirrors the traditional situation in science, where we are, to put it crudely, supposed to pay more attention to the reputation of the scientist than to the university they happen to work for or to the place their work is published. Scientists, like authors, are the brand. For Poster, digital cultural commodities bring even this identification of products with authors into question, offering us an increasing power to adapt cultural products and act not just as consumer, but as producer, creator, and user. According to Poster, material objects and digital cultural artifacts are inherently different in this regard: "The world of material commodities adheres to the brand and divides agents accordingly into producers and consumers. Not so in the domain of culture. Here the older legal and economic structures that insure commodification on the basis of authorship are disintegrating before our eyes" (Poster 2004: 421).

It seems that Poster thinks the importance of branding and of individual authors will diminish as cultural products become digitally available. This is an interesting provocation for cyberscience: will the advent of digital science repositories lead to the diminishment of the credibility cycle and of the individual scientist as epistemic subject? Will using a digital source of data rather than a published paper mean that traditional ways of judging worth and allocating credit are less important? Will retrieving information from digital taxonomic resources mean the end of citations to works of individual authors, and will it diminish the standing of the major systematics institutions? How will institutions and individuals adapt data-sharing practices to suit new regimes (Hilgartner and Brandt Rauf 1994)? In this chapter I assemble some observations on the prominent entities that inhabit the landscape of contemporary systematics and consider to what extent institutions and individuals are visible on this scene, showing that in a variety of ways digital resources are being designed to address the concerns of institutions about intellectual property and preserve a trail of credit back to the originators of data. The first section of this chapter makes a beginning on this project by looking at questions of visibility and topography, using a combination of interview data, Web surfing, and visualization of Web landscapes to find out what kinds of entities are involved in ICTs in contemporary systematics.

Landscaping Contemporary Systematics: Institutions and Initiatives

As I interviewed systematists involved in ICTs for this project, I kept alert to the landscapes that they mapped for me. As we talked about the issues I was hoping to understand, and as, in most cases, my interviewees began to warm to the project and get a feel for the kind of questions that interested me, they often began to recommend other places I should visit and people I should talk to. Ethnographers are used to informants recommending that they should be looking elsewhere for the real phenomenon (Rachel and Woolgar 1995), and it can be very telling when an informant suggests that someone else could be a better source than they. These recommendations help the ethnographer to map out what kinds of things their informants view as interesting, important, and noteworthy. Being recommended by an interviewee to go and talk to someone else is not (just) a bit of practical help that comes after the business of the interview proper, but is an important part of getting closer to someone's view of the world. Part of the work for this section was therefore done by talking to the people most concerned, working out what they saw as promising or interesting initiatives and where they felt that key things were happening, and taking note of their recommendations. I particularly mentioned the initiatives that figured in the Select Committee's report, to gain some triangulation on its portrayal of the contemporary systematics landscape.

In addition to the mapping that interviews provided, I also gained perspective on the field by surfing the Web. I started with institutions and initiatives that I knew, and spent much time trying to work out who was related to whom, which were the successful and unsuccessful initiatives, and who might be interesting interviewees to approach. Because my face-to-face interviews and my visits to institutions were inevitably geographically biased, I tried to offset this with my Web surfing, and to evaluate whether the view of systematics as cyberscience from the UK matched up with what other countries might see. When I came across acronyms that I did not recognize, I would try to track them down, and place them within my known universe of institutions, initiatives, governments, funding bodies, and nongovernmental organizations. I also noted names of individuals that cropped up repeatedly across initiatives. I developed, in short, my own personal archaeology (Rogers and Marres 2000) of the initiatives, institutions, and individuals of systematics as they were manifested on the World Wide Web.

Later this personal and largely informal version of the Web landscape was supplemented by a visualization technique provided by the TouchGraph Google Browser. This solution was chosen from a wide range of choices available for analyzing the topography of the Web. For example, numerous constituencies are interested in methods to assess how successful Web sites are in terms of the ease with which they can be found, which in turn often corresponds to the status that search engines accord them. In addition to commercial ways of exploring the Web focused on Web site link status, there are many other means to visualize the territory of the Web, as surveyed by Dodge and Kitchin (2000). There is also a wide variety of ways of visualizing and exploring the "social information spaces" of the Internet (Lueg and Fisher 2003). More specifically, analysis of hyperlinks has been an area of focused academic interest (Thelwall 2004). Most notable for the purposes of my research project is the speculation that hyperlinks to academic sites could act as indicators of reputation, analogous to citations in published literature: the webometric approach (Park and Thelwall 2003). A related approach takes analysis of interlinking as a way of exploring the network of organizational sites relating to selected issues, using visualizations of the density of hyperlinks as a way of exploring the polarization and segmentation of issue networks (Rogers and Marres 2000; Rogers 2002).

Another promising approach for studies of emerging landscapes in cyberspace is Web sphere analysis, allowing researchers progressively to identify a set of Web sites relating to an issue of concern, archiving them as they change over time, and analyzing the resulting temporally and spatially structured Web sphere (Foot et al. 2003; Schneider and Foot 2004).

These structured and visual ways of exploring the Web had strengths to recommend them. I could have conducted detailed hyperlink analyses of the various institutions and initiatives that made up my field of interest. Developing an independent measure of who was linked to whom, who most visible and who marginal or unsuccessful, is an appealing prospect, and one that I would not rule out for the future. The lure of Web-based data is that they are available for analysis, there to be harvested from the public domain, and ripe for the exploration of network structures (Beaulieu 2005). However, there are some caveats. Analyzing hyperlinks entails making numerous leaps of faith regarding the behavior of search engines, the appropriate units of analysis, and the meaning of linking practices: hyperlink analysis remains, as Park and Thelwall (2003) put it, a "complex and problematical tool." Just because Web links are collectable and analyzable does not mean that their interpretation is straightforward; triangulation between different methods will often be needed (Thelwall 2006). In science there may be disciplinary differences between linking practices, so any link-based analysis will have to make use of contextual interpretations (Harries et al. 2004).

"Objective" measures of success in the "public" domain of the Web also come with a certain amount of caution for an ethnographer wanting to retain some solidarity of understanding with mundane, everyday ways of experiencing the Web. While interviewing the authors of Web-based initiatives in systematics I came across few who said they had time and energy to look at their own log-files, let alone indulge in search engine optimization or analysis of their in-links. If I wanted to keep sympathy with their way of viewing the Web, I would substitute formal link analysis with a more in-depth approach to finding out how Web site owners experience the Web. This is not to say that a structured analysis of Web sites and hyperlinking is always antithetical to an ethnographic approach. Rogers (2002) argues for issue networks as useful tools to think with, and this is as applicable to ethnographers of the Internet as it is to organizations developing a Web presence. An in-depth hyperlink analysis would be very interesting, indeed, but not quite in the spirit of the current project or in line with what I found in the field. It would also, I feared, detract from the balance that I hoped to achieve between online and offline foci for ethnographic attention.

I decided therefore to continue with a more craft-based approach to exploration of Internet domains using readily available tools that other Web surfers or Web owners could use themselves, if only they had the time, in order to arrive at a visualization of the Web space that concerned me. Since Google is currently the dominant search engine, it seemed appropriate to deploy a visualization based on Google results.

The technology I chose to use in visualizing the landscape of institutions and initiatives was the TouchGraph Google Browser (http://www.touchgraph.com/TGGoogleBrowser.html/). This software is available online as a Java applet that takes a URL as input and returns a visual representation of a network of related sites centered on that URL, using Google's "similar pages" (also known as "related") facility to find related sites. The Google company does not reveal the precise algorithm behind the engine's judgment of related sites. The public account that Google gives of the "similar pages" function is as follows:

> When you click on the "Similar Pages" link for a search result, Google automatically scouts the web for pages that are related to this result.
>
> The Similar Pages feature can be used for many purposes. If you like a particular site's content, but wish it had more to say, Similar Pages can find sites with similar content with which you may be unfamiliar. If you are looking for product information, Similar Pages can find competitive information so you can make direct comparisons. If you are interested in researching a particular field, Similar Pages can help you find a large number of resources very quickly, without having to worry about selecting the right keywords. (Google 2004)

"Similar" obviously has a quite flexible meaning depending on the particular context. Without access to the algorithm used to generate the results, we cannot be certain what exactly constitutes Google's use of the term similar, but it could involve both shared keywords in the page content and shared linking structures, these being the main options for assessing similarity (Haveliwala et al. 2002; Thelwall and Wilkinson 2004). For sites to count as related in Google, they probably have to both share content and participate in networks of links as a result of being co-linked to by third party sites. The Google "related:" search returns, for those more prominent sites for which data is available, a list of ten related sites. The TouchGraph Google Browser piggybacks on this facility, generating the ten related sites for an initial seed URL, and the ten related sites for each of these ten, and then drawing the resulting network of related sites. Closely related sites cluster together in the visual display. Thanks to the importance of co-linking, the resulting graph can be interpreted as the upshot of third-party

Web site developers' judgments about which sites go together. For example, if Web sites listing useful botanical resources repeatedly included both Kew's ePIC portal and the Missouri Botanic Garden these two resources would appear closely clustered on the graph. This is certainly appealing from an ethnographic viewpoint, offering a chance to explore, albeit in aggregated and indirect fashion, the ways in which the territory is mapped out, from the perspective of those engaged in linking to the sites concerned and therefore generally practically involved in the field. It thus forms a nice complement to the approach I took in interviews, when I explored with people what they saw as significant initiatives and when they suggested to me other paths to pursue.

One of the appealing qualities of the TouchGraph Google Browser is its fluidity. The graphs that are generated grow on the screen as you watch, and then sway gently like fronds of seaweed as you pass the mouse cursor across them. Sections of the graph can be tweaked into new positions to explore the linking and clustering. New portions of graph altogether can be generated by double-clicking on a site to download a new set of related sites. Options are also available to limit the amount of the network shown by filtering out sites that have few links, focusing only on the most closely clustered results. Graphs do not sit as static representations of some preexisting structure, but are coaxed and crafted into existence. Another source of variability is the reliance on Google results, themselves changing as the crawlers do their work of exploring the Web. This is not an objective representation or a frozen analysis of an archive of hyperlinks. This fluidity has a strong lure for me as an ethnographer, as an invitation to think about exploration rather than representation, entering into a space rather than looking at a picture. The images I use in this chapter are screen shots of images crafted in this way, chosen to illustrate a particular point or grabbed at the moment when I saw an interesting connection. The static nature of the printed images makes it too easy to think of them as representative of the way the field is: their contingent qualities should, however, be borne in mind.

Accepting that we have no access to the algorithm that Google uses to generate related sites grants us a large amount of leeway in interpreting the results. I take them as provocations, constantly asking me to return to other sources of ethnographic insight, from interviews, from documents, and from reading and surfing Web sites more directly, asking whether these networks make sense as something I know from these other places. There is no single meaning to one of these network images, but it provides a handy object to think with. I use them here to illustrate a discussion of the entities involved in developing ICTs for systematics, and what their connections are with one another. This is necessarily a selective description. It is frankly impossible to do justice to the full range of initiatives involving digital information that are currently being developed across the international field of systematics. Instead of aiming for an exhaustive catalog, I focus instead on using selected examples to explore the kind of entities involved and how they relate to one another. My observations fall under two related headings: the structuring of the field in terms of initiatives and institutions; and the structuring influences afforded by technologies, by the various funding bodies and schemes, and by the ongoing process of standards setting.

New Initiatives and Old Institutions

One of the first observations is that the traditional institutions of systematics, the natural history museums and botanic gardens, are key players in the field of contemporary initiatives. As described in chapter 4, virtual developments in systematics are thoroughly rooted in and sustained by its material culture. Initiatives to make information available on the Internet often begin with the existing institutions and involve the digitization of specimen information or nomenclatural resources. Viewed using the TouchGraph Google Browser, however, the results are a little surprising. Figure 6.1 is a representation of the related sites network for the Natural History Museum, selected to show only the most linked sites. This is a territory populated by peer institutions, including other British national museums and other science and natural history museums overseas, such as the Museum Nationale d'Histoire Naturelle in Paris, a number of museums in the United States, and the Deutsches Museum in Munich.

This is not, however, how most of the systematists working in the museum would view their relevant networks. "Museum worker" is an important identity, but it is not the salient one for most of the activities that the systematists employed by museums are engaged in. The institutional network is certainly not a fair representation of how the museum relates to other institutions as far as the provision of more specific online resources is concerned. If we draw the network of related sites centered on one initiative at the museum we find a very different structure. Figure 6.2 shows the network of related sites for the LepIndex project, comprising a digitized version of a card index detailing global Lepidoptera names. This network consists of a set of other databases focused on Lepidoptera in various countries. The Caterpillar Hostplants database is also located at the Natural History Museum, as is the UK Buttterflies and Moths site, while Moths of

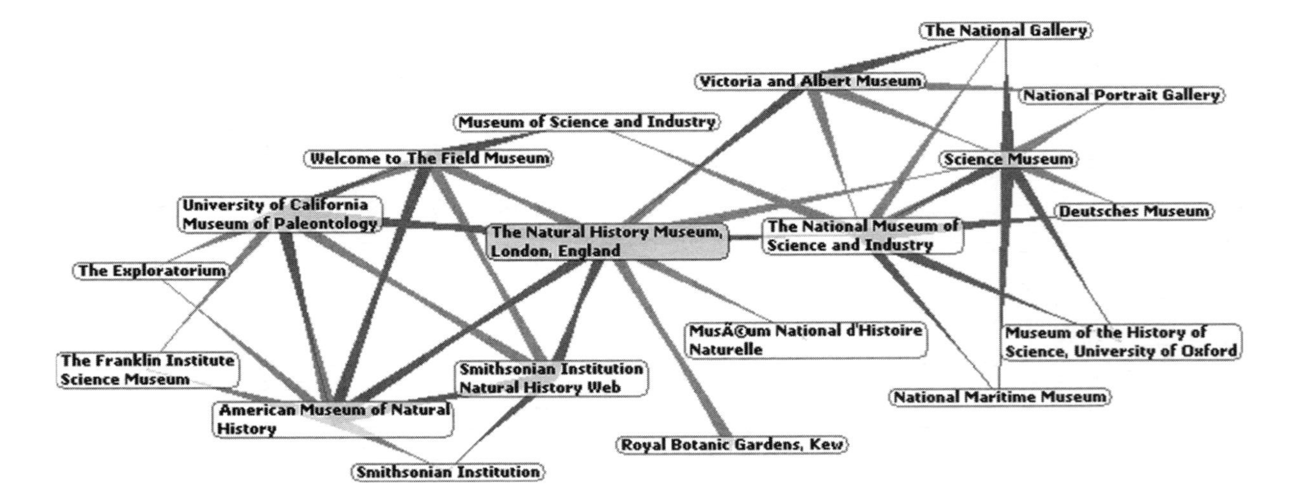

Figure 6.1
Touch Graph Google Browser network for the Natural History Museum, London. (Note: as "found" artifacts, the Touch Graph images have been reproduced exactly as they were originally captured, including, as in this case, misrenderings of diacritical marks.)

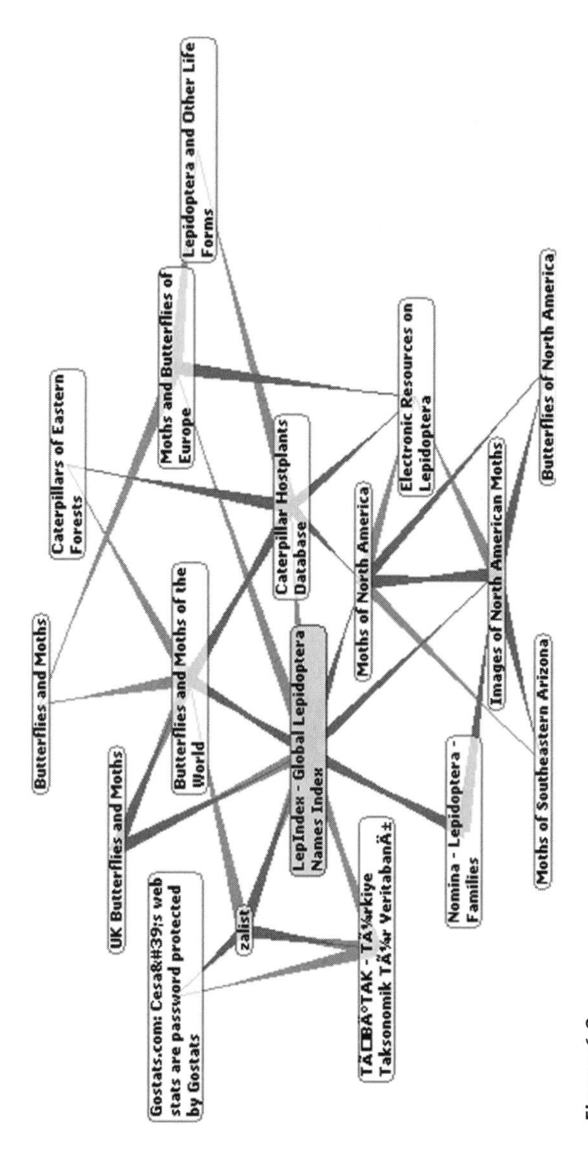

Figure 6.2
Touch Graph Google Browser network for the LepIndex Global Lepidoptera Names Index.

North America is a site owned by the US Geological Survey, TUBITAK is a taxonomic database for Turkey, and Lepidoptera and Other Life Forms is a Finnish resource. This network is then also an international field, but related to a specific topic rather than to similar kinds of institution: in fact, on following the links to each of these resources one has to dig quite a way in some cases, to identify the institutional affiliation. Although institutions may contain many different resources, then, the set of connections that a particular resource participates in may be very different from those that are relevant for the institution as a whole. Clearly certain initiatives are seen as part of a peer network in which being a museum is less relevant, and being a resource relating to a particular group of organisms more so.

A set of relationships very significant for the LepIndex project but not visible in its network of related sites is that of the collaborative links and funding sources that went into its production. In chapter 4 I described the ingenious solution developed in the LepIndex project to scan a card index of nomenclatural information in order to make it available as an online database. This project was a collaboration between staff at the Natural History Museum in London (including entomologists and Web technologists) and staff from the Department of Electronic Systems Engineering at the University of Essex. The project took a card index developed over decades of painstaking manual updating and invented a way to turn the information it contained into a searchable Web-based database. Crucially, the need for ingenuity in carrying out the project made a source of funding available, via the joint Bioinformatics Initiative between the Biotechnology and Biological Sciences Research Council and the Engineering and Physical Sciences Research Council. Were it not possible to present the project as technologically innovative, it is unlikely that this source of funding would have been available to make the project a reality. The data set was also portrayable as a comprehensive and authoritative nomenclatural resource, which made further sources of funding available, both internally through the museum's strategic initiatives to enhance its Web-based offerings, and externally through the ENBI (European Network for Biodiversity Information) and Species 2000 Europa (two interinstitutional networks, of which more later). The LepIndex project therefore figured as a fundable initiative both in virtue of how it was carried out and the strategic significance of the data it contained. The network of related sites demonstrates some key users of this resource, but does not map the funding bodies who contributed to it or the two collaborating institutions that created it.

A similar divergence between institutional networks and initiative-based networks operates for the Royal Botanic Gardens at Kew. The network of

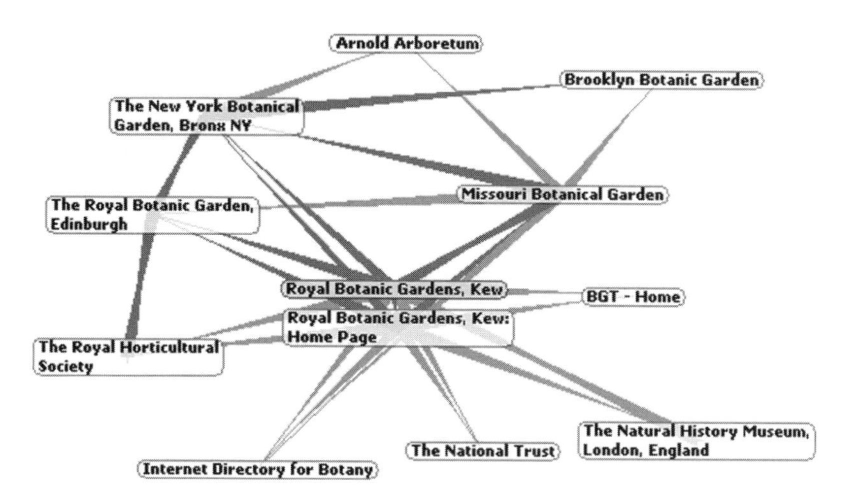

Figure 6.3
Touch Graph Google Browser network for the Royal Botanic Gardens, Kew.

related sites for the institution, shown in figure 6.3, is slightly complicated in that two URLs for the main site are in use (www.rbgkew.org.uk and www.kew.org), but it shows the same general pattern of a set of peer institutions, consisting of museums and botanic gardens both in the UK and internationally (BGT being the Botanic Gardens Trust in Sydney, Australia). The only noninstitutional site visible in this network is the Internet Directory for Botany, an index maintained by botanists from several institutions in the United States, Canada, and Finland.

The Royal Botanic Gardens at Kew has promoted its data holdings via a common gateway to all of its online databases, the ePIC or Electronic Plant Information Centre. This gateway was developed through funding from the Capital Modernisation Fund, a government initiative intended generally to promote improvement of public service delivery and not confined to information technology–related initiatives. Building the gateway as a specifically named entity, rather than leaving individual holdings as a collection of diverse data resources, made it plausibly presentable as an appropriate initiative under this funding rubric, with the goal of making plant information resources available to a wider audience of systematists and users in the academic community and beyond. The ePIC network of related sites, portrayed in figure 6.4, is quite different from the overall institutional network. This is also very much a peer network, but here many of the peers are other databases and online plant resources, including the Flora of North America, the Missouri Botanic Garden's W^{3}TROPICOS nomenclatural database, the Index

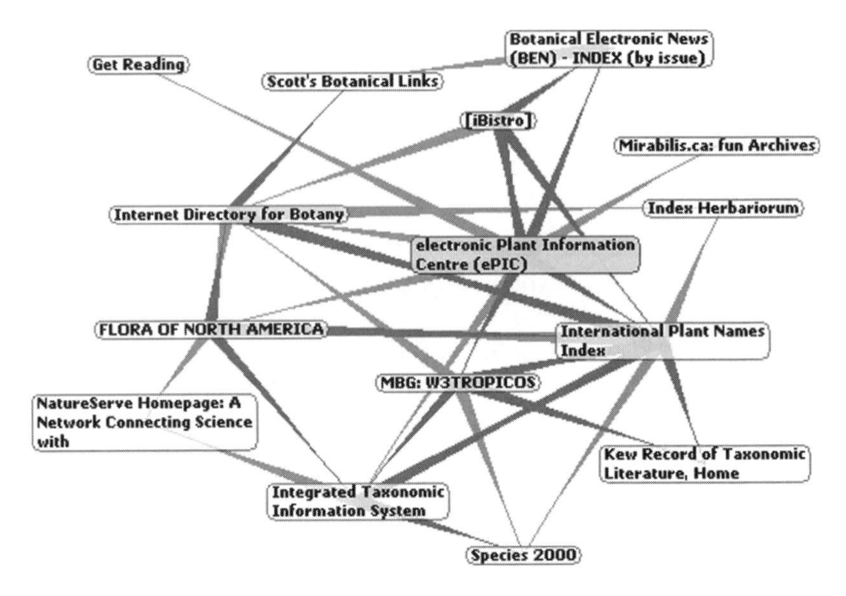

Figure 6.4
Touch Graph Google Browser network for the ePIC Electronic Plant Information Centre at the Royal Botanic Gardens, Kew.

Herbariorum of the New York Botanic Garden and the International Association for Plant Taxonomy, the NatureServe conservation-related databases, and Kew's own Kew Record and iBISTRO, the Kew library database. The network also features some index sites: the Internet Directory for Botany, Scott's Botanical Links, and a news service, the Botanical Electronic News.

As suggested earlier many Web site owners do not prioritize the assessment of their site's success; but there are instances where for institutional purposes, or to satisfy funding bodies, it is deemed appropriate to invest some time in analyzing who is using a site. Such an evaluation has been done for the ePIC gateway, since it represents a new departure for the institution in its presentation of digital resources and is a focus of particular institutional attention: targets for use of Web resources, including ePIC, are set as part of the institution's corporate plan (Royal Botanic Gardens Kew 2004). Analysis done by the Web site developer in the early days of the gateway using log-files showed that the predominant usership came from the developed world, although a minor but politically and institutionally highly significant number came from Africa and tropical Asia. Log-files are, however, often as frustrating as they are informative for Web service designers, giving information about patterns of data access but little means to assess the motivations and experiences of the people behind the queries. The ePIC example is likely to be typical of the experience of Web service developers, who will have a quantitative sense of levels of success of their sites, but be restricted in their access to what this means in terms of the working practices of their users. There is little direct feedback from users about their levels of satisfaction with services, and it can be immensely frustrating to see patterns of hits emerging without being able to work out what this means or how, as is so often important for service providers, to increase the number of hits received. Even comparing performance with other similar sites is problematic, since hits can be counted in different ways, and the same data is often accessible via different routes.

An important entity on the ePIC network, and one that considerably predates it, is IPNI, the International Plant Names Index (Lughadha 2004). This joint venture between Kew, the Harvard University Herbaria, and the Australian National Herbarium combined the largely complementary nomenclatural resources of the three institutions to make one relatively complete index. This interinstitutional project was developed to make a more complete resource than could have been achieved by the partners individually and came with a major rationale, at least for Kew and Harvard, in terms of reducing their duplication of effort in indexing the taxonomic literature. Such collaborative efforts have the dual rationale for institutions of increasing the effectiveness of information services for users of systematic information and of being seen to do so: it is important for contemporary systematics institutions to demonstrate that they are part of a field that is interested in becoming more efficient and moving toward a common goal. Being politically high profile entails becoming accountable for efficiency and effectiveness. The IPNI network of related sites, shown in figure 6.5, includes several entities we have already encountered and adds a few more to the mix. The International Code of Botanical Nomenclature is related to IPNI not as another information technology–related initiative, but rather through its concern more broadly with the maintenance of a stable nomenclatural system. Its occurrence in the network reminds us that projects like IPNI are not, after all, to be reduced to strategic responses to a trendy technology. Instead, or as well, they are part of ongoing efforts in systematics, going back through centuries, to work out how to give stable and usable names to groups of living organisms.

Other links in the IPNI related sites network are more obviously part of the cyberscience turn in systematics, most notably the meta-initiatives Species 2000 and GBIF, which I will discuss later. The Expert Center for Taxonomic Identification (ETI), based in Amsterdam, is a UNESCO-related initiative with

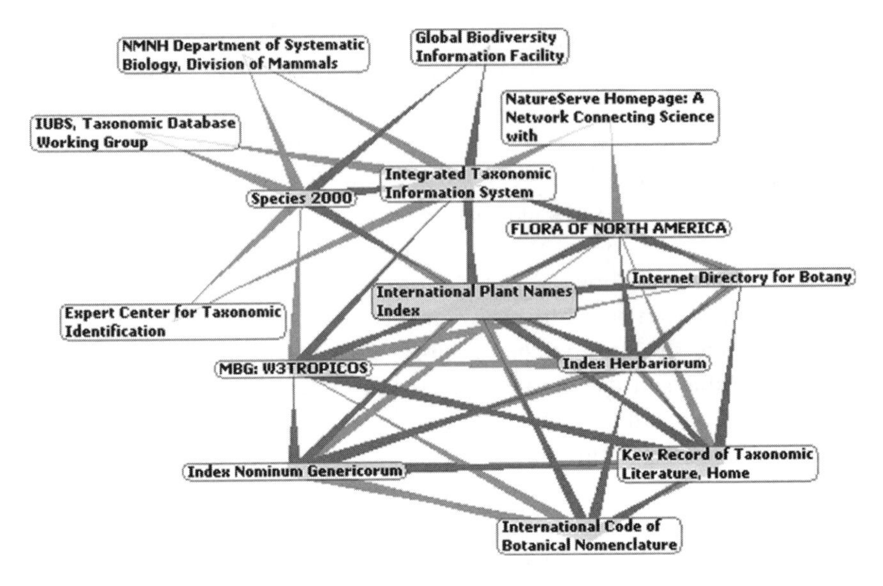

Figure 6.5
Touch Graph Google Browser network for the International Plant Names Index.

a mission "to develop and produce scientific and educational computer-aided information systems, to improve the general access to and promote the broad use of taxonomic and biodiversity knowledge worldwide" (Expert Center for Taxonomic Information n.d.). In practice this means a focus on developing and distributing software for computer-based identification and products based on this software, plus a small set of databases including a global directory of taxonomists. ETI is indirectly related to IPNI via Species 2000 and ITIS, both of which aim to gather authoritative name listings for particular groups of organisms, striving toward the complete catalog of recognized names that would provide a gateway for searches on biodiversity and conservation-related information. Also linked in this way, somewhat cryptically, the node labeled NMNH Department of Systematic Biology, Division of Mammals, is in fact an authoritative, collaboratively produced list of the mammal species of the world. IPNI participates in a network along with other site related by their status as authoritative collators of names in a particular group of organisms. They are colocated in a network not because they provide information about the same thing, but because they represent similar kinds of activity.

We can see, then, that the traditional institutions of systematics are visible on the Web, and that they participate in mutually visible peer networks. It would be hard to conceive of a contemporary museum in the UK, at least one with governmental funding, that finds it unnecessary to have a Web presence. Little of the scientific collaboration is, however, actually visible within the networks drawn at the level of institutions. For this we have to look to more specific projects. Within the institutions there are many individual initiatives, relating to a particular group of organisms or geographic area or a key historical collection or bibliographic resource. These initiatives come about as a funding source is brought together with an available data source.

As discussed in chapter 4, digitization is an occasion for prioritization, and decisions to make a particular resource available online depend on being able to position it as a fundable project. The dual interest that systematics institutions display in moving toward more effective information services and in being seen as keen to do so is a consequence of a constrained funding climate within a highly politicized domain. The work of systematics institutions has become increasingly significant on a global stage, as signatories to the Convention on Biological Diversity are required to demonstrate their compliance with its provisions. On a more basic level, the Convention on Biological Diversity has been interpreted as increasing the impetus toward computerized means for tracking the movement of specimens on loan, as institutions are required to keep detailed records. On both grand and more mundane levels, then, institutions are participating in computerization initiatives: because it makes practical sense to do so, because they have to be seen to do so, and because access to funding often requires that they do so.

Tempering this overall will to digitize is the fact that relatively small amounts of funding tend to be available for relatively short periods of time, producing a patchy coverage of overall specimen holdings and some isolated resources: a certain amount of "cherry-picking" involves the kind of prioritization decisions described in chapter 4. In the major institutions, many of the isolated initiatives have been the work of individual members of staff developing an idea, pursuing a specialism, or championing a particular resource. Piecemeal approaches have been successful in digitizing selected bodies of data. They have also been important as showcases and proofs of concept, showing that collections and documentation have been digitized, and helping to create a climate across the systematics community that begins to expect important data sets to be made available online. As discussed in chapter 5, the Taxacom list has provided a useful route for publicizing resources and for creating a climate of expectation around digitization. At the same time, pioneer projects provide occasions for learning about the true costs of digitization, and for propagating practical advice about how to manage such projects (Smith et al. 2003).

The piecemeal approach has also brought some problems, particularly in terms of incompatible software and data schemas. Considerable effort has lately been focused on promoting use of standard data fields and developing gateways to allow searching across different data sets, such as the ePIC system at the Royal Botanic Gardens, Kew. At the Natural History Museum, moves to adopt a common software for cataloging across the different departments have become realistic only relatively recently, with the exploration of a commercially available piece of software for collection cataloging: until this point solutions were developed locally according to the needs of the particular department and its understanding of what it is important to record about specimens. Systematists have for many years developed their own personal databases to keep records of the specimens they have examined. These individual databases have recently been recognized as a valuable resource, but there can be considerable problems in making them more widely available by adding them to the overall catalog if they have not been designed in the right way. One way of promoting standardization short of absolute fiat has been employed at the Kew herbarium: researchers are free to construct their own databases, but may use Kew barcode labels for specimens only if their core database fields are compliant with the overall Kew catalog.

If there is a problem of compatibility within data sets on an institutional scale, the problem is magnified considerably when the goal is to provide access to information across institutions and internationally. Beyond the initiatives of individual institutions and tightly conceived collaborative projects between small numbers of partners, there are several initiatives on a wider international scale which aim to provide for interoperability of databases. Within European institutions there has been a series of related efforts to bring together collections databases to provide Europe-wide access to information on specimens. The European Network for Biodiversity Information, ENBI, was funded as a thematic network under the European Union Framework 5 funding program, acting formally as the European response to GBIF. As a thematic network this program did not conduct its own research, but instead acted as a broker between participants, many of whom were involved in responses to GBIF at a national level, and considered issues relating to coordination, maintenance and enhancement of databases, data integration, interoperability and analysis, and products and e-services. The products were reports, Web-based disseminations, and expert forums.

Another Euopean level initiative was BioCISE (Resource Identification for a Biological Collection Information Service in Europe), a Concerted Action Project funded by the European Commission. BioCISE was largely a

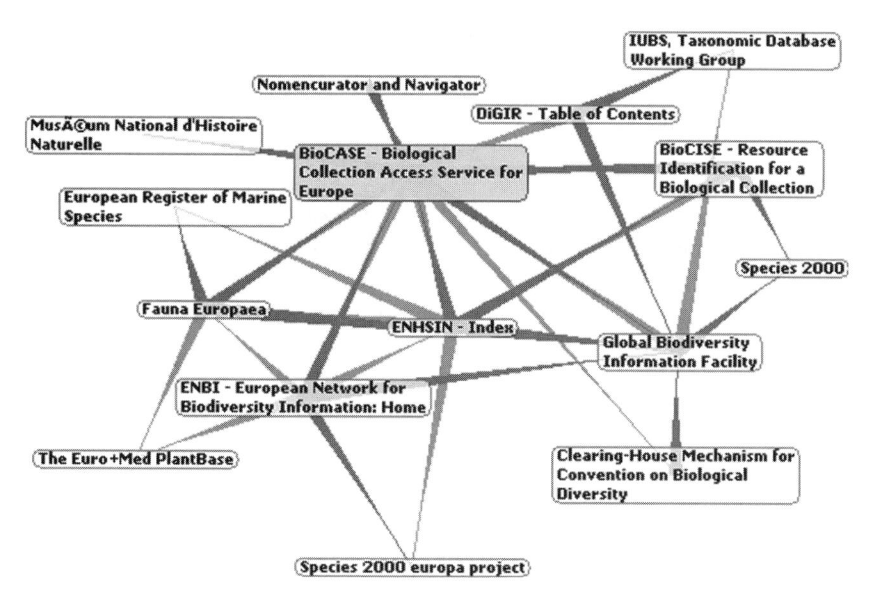

Figure 6.6
Touch Graph Google Browser network for the BioCASE (Biological Collection Access Service for Europe) project.

scoping and feasibility study and was succeeded by BioCASE (Biological Collection Access Service for Europe). Also operating in similar territory and feeding into BioCASE was ENHSIN, The European Natural History Specimen Information Network, funded under Framework 5. I show here the network of related sites for just one of these initiatives, BioCASE, in figure 6.6. Other European-level projects appear here, ENBI, ENHSIN, BioCISE, and the Species 2000 Europa project, together with the meta-initiatives GBIF and Species 2000. Another kind of entity also makes an appearance in this network: DIGIR (Distributed Generic Info Retrieval) and TDWG (Taxonomic Databases Working Group) are both concerned with data standards. These issues arise most starkly when attempts are made to link diverse databases, as the European projects often do. It is therefore apt that the standard-setting bodies should appear in the sphere of European projects. I will be discussing them in more detail in the next section.

At the European level, the systematics community demonstrates considerable activity in promoting access to biodiversity information and interoperability of databases. The names of projects vary: some interviewees told me that what I thought was an explosion of branding in e-science projects generally could be understood as effects of the EU focus on project acronyms.

For each new funding cycle a set of partners must be assembled, a project formulated as a set of distinct "work packages," and a new name and acronym agreed on for the project. It has become accepted practice for this kind of pan-European project to develop an approximation to a "corporate identity" including a Web site and published reports. The networks of partners involved in an EU program are often large and can be diffuse and unwieldy. A corporate identity is useful both internally and externally in these circumstances, helping to convince and remind partners that they are engaged in a real endeavor, and at the same time presenting a coherent face for what often does not feel like a very coherent set of activities.

European initiatives need to be viewed in the context of concerns to develop a specifically European dimension to research, connecting together nationally based research groups and institutions in a consciously cross-national fashion. Whereas the majority of the funding that systematics institutions receive tends to be nationally based, the advent of European funding has provided the rationale, or the impetus for the specification of networks on this basis. Again, the situation is a complex one: a source of funding such as this provides a rationale for forming particular sets of collaborations, and it also provides a route for exploring avenues that might already be on the agenda, albeit in a particular context. In this case, the question of interlinking databases is high on the agenda throughout systematics, and the European funding provides a site to work through these issues and a set of motivated partners to collaborate with.

For institutions located within the countries of the European Union, the availability of funding for networks specifically to introduce a pan-European dimension has provided an additional stimulus for attempts to form interoperable and cross-searchable databases. Although there has been a European response to the Convention on Biological Diversity and ENBI acts as the European response to GBIF, in most cases a response has also been organized on a national level since it is national governments that sign up in both instances. In the UK, the National Biodiversity Network (NBN) brings together a range of conservation organizations, including the National Federation for Biological Recording (NFBR) and the UK Biodiversity Action Plan (UK BAP). Present in this network of related sites (figure 6.7) is a mixture of users and providers of biodiversity information. The systematics institutions are notably absent in relation to the national initiative, much of the response to the Convention on Biological Diversity in terms of providing nomenclatural information and specimen databases having happened either at institutional level or international level and bypassed the national.

Thus far I have deferred talking in any detail about what I have referred to as meta-initiatives. In this category I place Species 2000, ITIS (Integrated

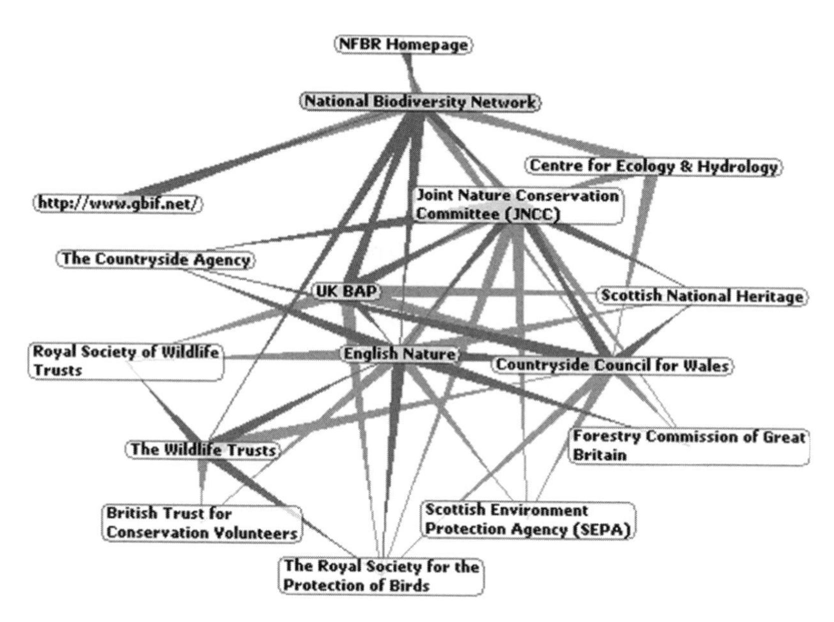

Figure 6.7
Touch Graph Google Browser network for the Natural Biodiversity Network.

Taxonomic Information System), and AllSpecies, together with GBIF in a somewhat special category all of its own as a meta-meta-initiative. These projects participate in a vision to raise biodiversity knowledge and information availability to a new level, aiming to achieve unified access to biodiversity information held in diverse places and forms, and to produce a unified catalog of all known living organisms. This kind of initiative is newsworthy, in scientific circles at least: assessment of the goals, relationships, and likelihood of success formed the subject of articles and correspondence in *Nature* (Bisby et al. 2002; Gewin 2002). In the next major section of this chapter I talk in more detail about Species 2000 and the mechanism for arriving at a consensus catalog which its predecessor, ILDIS, developed, and about some of the processes of enrolling systematists and identifying users that this entailed. I will be using this project to explore the complex dynamics happening within an initiative that are hidden in a survey at the level of initiatives and institutions. I will leave those complexities for now, and focus on the superficial relationship between meta-initiatives and institutions. In this regard it is important to note that none of these meta-initiatives is the domain of a single major systematics institution. Species 2000 is based at the University of Reading, but describes itself as a "federation" of database organizations. ITIS is based at the National Museum of Natural History in Washington, DC, but is formally a cooperative project conducted on behalf of several federal agencies of the US who have signed a memorandum of understanding.

AllSpecies had ambitious goals to raise awareness and funding for completing the cataloging of the world's flora and fauna, going beyond the work of ITIS and Species 2000 on existing taxonomic knowledge to focus on the additional work needed to explore the world's hitherto unknown biodiversity and aiming at a Web page for every species. AllSpecies explicitly positioned itself as an outsider, as its Web site proclaims: "As a new and innocent player in the taxonomy game who questions the rules, traditions, and pecking orders, ALL challenges the attitudes, mindsets, and perspectives of the players" (All Species Foundation n.d.). Fundraising was insufficient to allow the aims of AllSpecies to be turned into a concerted substantive effort: it persists as a "decentralized" effort with a public face provided by its Web site. From the outset AllSpecies was controversial, described in *Science* as "raising both the hopes and hackles of taxonomists" (Lawler 2001). For many of the people I interviewed, AllSpecies was seen as an example of a project that took the wrong tack, in having unreasonable expectations about the readiness of institutions to donate data sets, or in trying to build afresh without taking due account of what others were doing, or in being overfocused on corporate funding at the expense of substance. I am in no position to make assessments of how fair these comments were. They are interesting here largely in what they say about people's judgment of what a successful initiative would have to do. It is clear that there is a strong consciousness about the ownership of data, or at least the question of giving and continuing to give due credit to data providers, and about the need to work together with other initiatives and institutions. This is not a field in which "wiping the slate clean" is a popular sentiment. The initiatives that are thought to have a chance of success are those that allow for existing projects, institutions, and initiatives to come together, and which allow time for standards to be developed, differences incorporated, and concerns to be worked through. Initiatives must be fundable, without participants feeling that they are being driven by funding. Successful projects, as I will show in the next major section of this chapter, have to involve highly astute practical politicians. On that criterion AllSpecies was insufficiently successful, at least in the eyes of the various people I spoke with.

Finally, for this section, I turn to GBIF (featured in the TDWG network, figure 6.8). In chapter 3 I discussed GBIF as an evocative brand. Here I turn explicitly to its role in the landscape of initiatives. If the other initiatives and meta-initiatives are sites of practical politics, then GBIF has taken this to a new level of complexity. GBIF was set up by the OECD Megascience

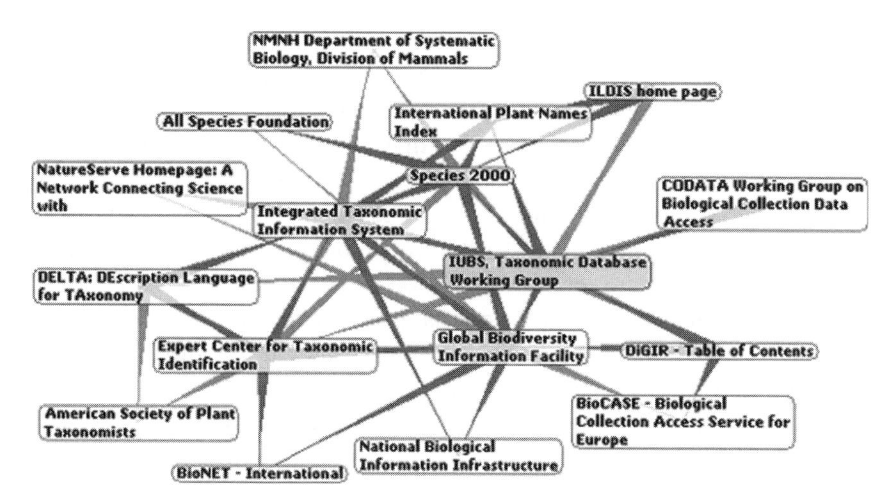

Figure 6.8
Touch Graph Google Browser network for the Taxonomic Databases Working Group.

Forum Working Group on Biological Informatics and formally launched in 2001. The Memorandum of Understanding that participants are required to sign sets out the scope and significance of its endeavors and ties these in with biodiversity concerns thus:

> The signers of this non-binding Memorandum of Understanding, being countries, economies, or inter-governmental organisations, or entities designated by countries, economies, or inter-governmental organisations, have decided that a co-ordinated international scientific effort is needed to enable users throughout the world to discover and put to use vast quantities of global biodiversity data, thereby advancing scientific research in many disciplines, promoting technological and sustainable development, facilitating the equitable sharing of the benefits of biodiversity, and enhancing the quality of life of members of society. The importance of making biodiversity data openly available to all countries and individuals is underscored by various international agreements, especially the Convention on Biological Diversity. (Global Biodiversity Information Facility 2000)

GBIF has grand aims to provide common portals to biodiversity information wherever it may be. To achieve this it needs authoritative species listings, for which it calls on Species 2000 and ITIS; it needs standards for data description and exchange, to which end it supports TDWG; and it needs information to be gathered in from wherever it is held, for which it needs a system of national nodes that contribute data. It also needs exemplars for the kinds of new project that are envisaged to capitalize on the availability of online information, using digitized resources to explore biodiversity, map and visualize it in new ways and deploy data-mining techniques to answer new kinds of questions. It is this kind of novel possibility enacted around unified access to dispersed digital resources that provides the rationale, in public rhetoric, for the meta-initiatives to proceed (Edwards, Lane, and Nielsen 2000). GBIF also aims to stimulate appropriate developments at a more "grassroots" level, through organization of competitions to allocate funding to specific digitization and database projects.

If we survey the various projects that populate the field of systematics as cyberscience, it quickly becomes clear that many of the things that GBIF aims for are already going on, albeit in piecemeal fashion, and that much of this activity predates GBIF. GBIF has taken great care to take account of existing initiatives and aimed not to reinvent them. One function of GBIF is, according to an early description, to bring together projects that might otherwise be in competition to encourage compatible approaches to achieving the overall goal (ibid.). It is significant that ITIS and Species 2000, instituted separately with the goal of producing lists of all species, have come together to provide nomenclatural information for GBIF. Also, GBIF takes considerable care to allow participants to feel they have control over their own intellectual property, through its focus on a common gateway to diverse distributed databases. Distributed databases have the advantage of being maintained up-to-date at their site of origin, but are also popular in that they allow any data accessed, through whatever portal, to be clearly marked with the institution that provided it. They can, therefore, to a large extent, deal with institutional concerns about giving data away, although they do not of course address some of the more complex concerns discussed in chapter 4, such as the possibilities of propagating erroneous information, endangering rare plants by making information available to the wrong people, or infringing on the tacit intellectual property conventions that operate when people lodge specimens in collections. These concerns still have to be dealt with by individual institutions developing policies on information disclosure to suit their own circumstances, although these individual arrangements will in turn be shaped by the practices that become conventional across the community.

GBIF has won a lot of awareness and acceptance, although not universal acclaim. As I interviewed systematists who were involved in various initiatives at systematics institutions, I also noted their reactions when I introduced the topic of GBIF. Wry smiles, laughter, or rolled eyes were common. The overall reaction was not one of unalloyed confidence in the new project. One systematist expressed a hope that it would not be "just another talking shop." Another spoke of it, somewhat ironically, as the "portal to end all portals." Most agreed that the signs were positive, that key parties were involved, and that sensible actions were being undertaken. There was certainly an acceptance that the ultimate goal would be worthwhile, although there were some doubts about who, exactly, was out there waiting to use such a portal and how they would use it. The general response was, however, overwhelmingly wary, perhaps understandably so from a community that has become tired of initiatives and knows, from bitter experience, how carefully crafted cooperation and consensus have to be. It is significant, though, that as the GBIF Memorandum of Understanding makes clear, the provision of online information relating to biodiversity and systematics is seen as a scientific task. It might have been argued that the task was one of computing or information science. There are, of course, representatives of these communities involved in projects, and the task is not one for which skills are already available within the systematics community. Nonetheless, the membership and participant lists of the various initiatives make clear that the existing systematics institutions have taken on the major burden of defining the tasks to be done and carrying them out.

The proliferation of initiatives and acronyms is dizzying, each being necessary to bring together a particular subset of partners, to focus a specific effort, to enroll a given funding source, or to appeal to a defined set of users. Nonetheless, the key players remain the same, and the key systematics institutions remain organized, as ever, in peer networks with high levels of mutual visibility. New forms of initiative act to connect them in new and more intricate ways. The global significance of biodiversity information together with the expectations of information availability promoted by new technologies have both made new funding available for systematics and placed new burdens on it. The old institutions of systematics have risen to the challenge, but in ways that are understandably cautious of giving away what resources they have. This caution is prominent in the discussions over intellectual property that form a part of the major network initiatives, and in the strong preference for information retrieval via distributed databases rather than a once-and-for-all amalgamation of diverse datasets into single resources.

Structuring Influences, and the Dance of Initiatives

The previous section showed that institutions provide considerable structure to the field of contemporary systematics, although new initiatives and projects are becoming key players and linking institutions in new ways. Running through the sometimes dizzying parade of interlinked initiatives and acronyms, and the recurring institutional partners in different configurations, was another set of structuring influences. These influences act to stimulate activities in the provision of digital systematic information and also to limit the scope of possible responses to these pressures. In this section I focus on some significant structuring influences that, in addition to the preexisting major systematics institutions, act to shape the field of systematics as cyberscience. These influences are: the shape and timescale of funding regimes; the Convention on Biological Diversity and the view of systematics itself as infrastructure; and the role of the Internet as an infrastructure for systematics. I will discuss each in turn, for the extent to which it shapes possibilities and the levels of influence it has on the outcomes. I will then go on to talk about the development of new infrastructures for the sharing of systematics information via standards setting.

I have already alluded several times to the role of funding regimes in making particular projects possible and in promoting specific kinds of projects. The sources of funding for work on ICTs are very diverse. Systematics institutions have had to become highly creative fundraisers in order to survive what are, in the UK at least, felt to be years of underfunding, as documented by the Lords Select Committee report (Select Committee on Science and Technology 2002a). The 2004 annual report of the Natural History Museum in London lists grants for research from an array of UK funding bodies and government initiatives (the Darwin Initiative, Biotechnology and Biological Sciences Research Council, Natural Environment Research Council, Wellcome Trust, The Health Foundation, Royal Society, Department of Trade and Industry) plus five different EU schemes, all commencing in that year. For 2003–2004, the Royal Botanic Gardens at Kew report project funding from a similarly numerous array of sources including government departments and initiatives (the Department for the Environment, Darwin Initiative, Foreign and Commonwealth Office, English Nature), charitable foundations (Andrew W. Mellon Foundation, Millennium Commission, Charles Wolfson Charitable Trust, Nestlé Charitable Trust), and other nonprofit organizations concerned with biodiversity (Global Biodiversity Information Fund, International Plant Genetic Resources Institute, and Earthwatch).

The diversity of funding sources speaks to the health of the field and its ability to enroll different audiences, but also provides some tension through the resulting diverse structures of accountability into which it ties systematists and the varying prioritizations that different audiences promote. This is not to say necessarily that funding bodies are driving the priorities of systematics institutions: this would be viewed as unacceptable by the institutions and by the individuals involved. Nonetheless, it is clear that what can be done in the direction of digitizing collections and resources will be focused in part on what can be presented as a discrete fundable project. Digitizing has become part of the core work of curating, but even to catalog entire collections on the basis of the available core effort would take decades or centuries: to move faster would require considerable additional investment (Blackmore 1996). Funding that is available for specific projects, and for participation as members in a transinstitutional initiatives, thus becomes an important part of the operations of systematics institutions.

On another scale altogether, funding regimes are important in structuring projects, and in structuring institutional involvement in projects. The prime example of this form of structuring is the work package of European funding regimes. These provide projects with a structure within a structure, allowing for smaller groups of participants to come together on specified areas within the project. In this way, institutions can become specialists, and individuals can pursue a particular area of expertise while ostensibly being involved in one grand project. Membership and participation are variable acts that can mean quite different things. The formulation of European funding in terms of work packages and the allocation of those work packages to subsets of partners allow for some to take a more prominent role than others, and for those with a particular agenda or specialization to focus on that. Work packages allow areas of autonomy to develop within the project. In this way, a smaller institution can participate meaningfully without having to be involved in the whole scope of a large project. A larger institution can take a lead role in a work package and develop a greater impact on the overall direction. The European funding regime provides just one example of the way in which the requirements of a particular scheme can shape the form participation takes, offering a structure that is then adopted and made meaningful in particular circumstances.

In chapter 4 I wrote about the specimen collections held by systematics institutions as a virtual instrument for exploring the variation of living organisms. Another way of putting this might have been to describe the collections, and the conventions for loaning specimens and welcoming visitors to collections, as an infrastructure for systematics. Virtual initiatives are being interwoven with the material components to extend the reach of this infrastructure, promoting its accessibility and attempting to overcome what are seen as the shortcomings of material specimens by augmenting them with virtual ones. This perspective on the network of systematics institutions as providers of an infrastructure for study of biodiversity has received endorsement through concrete funding. The SYNTHESYS project, funded under the European Framework 6 Integrated Infrastructure Initiative, both provided funding for systematists to visit selected institutions from the network of partners, and also aimed to move toward seamless "virtual" museum services through common standards and policies and by joining up databases. Many of these activities might also feature in other initiatives, but are brought together here specifically under a banner of creating an infrastructure for the study of systematics, building on the traditions of specimen loan and acceptance of visitors that have previously constituted the systematics institutions as a scientific instrument for the study of biological diversity.

While discussing structures and structuring influences, it is important also to note that the products of taxonomy themselves can count as infrastructure. In 1991, as part of the research for my DPhil thesis on the issue of nomenclatural instability, I collected definitions of taxonomy that I found in textbooks and other public pronouncements on the discipline. I divided the definitions I found into three groupings, on the basis of the purposes they gave for constructing taxonomies, and the principles for judging a good taxonomy. A significant but numerically small category I termed the "selfish" definitions, which suggested that classification was an intellectually satisfying activity, done for one's own appreciation and that of one's immediate peers. A much more prominent category in frequency of occurrence was the "scientific" definition, which stressed that the purpose of a classification was to embody hypotheses and to provide for inductive generalizations. More prosaically, the category of definition I termed "marketplace" stressed the utility value of taxonomy and its practical function of providing a system of unambiguous names for the rest of biology to build on, to provide access to the literature and offer a basis for identification. This kind of definition positions taxonomy as an aid to information retrieval. Marketplace definitions were, in my unsystematic sample, most common of all. A number of authors combined marketplace and scientific definitions, concluding that it is by constructing classifications with a scientific basis that taxonomists best serve their goal of providing a stable and meaningful information retrieval system for biologists.

There have always been tensions in positioning taxonomy between service and science and in finding approaches to nomenclature and classification that meet both requirements: this is the "duality at the heart of taxonomy" (Thiele and Yeates 2002: 337). The issue is important, both for the image of the discipline and for access to funding. Godfray and Knapp, in the introduction to their collection of papers on "taxonomy for the twenty-first century," express the tension in eloquent terms: "Taxonomy is a proud and independent science whose achievements rank as some of the outstanding successes of modern science. It is not a technical support service for the rest of biology, but a science that advances through testing hypotheses about taxon status and phylogenetic relationships. Yet taxonomy stands alone, divorced from the rest of biology, at its peril" (Godfray and Knapp 2004: 560). They go on to make the connection with funding yet more clear: "taxonomy needs a strong and vocal constituency rooting for it in the increasingly cut-throat competition for research funds" (ibid.). This is a clear example of a "marketplace" approach to taxonomy, yet at the same time one that holds to a notion of scientific independence.

It has at times been controversial to stress the service aspect of taxonomy, for systematists who were anxious to promote their activities as basic science. Projects that aim to produce unified catalogs from smaller-scale dispersed resources have in the past claimed that they are rejected unfairly by scientific funding bodies who either misunderstand the scientific nature of the work involved, or underestimate the important value of such resources (Sun 1987). It is also said that projects to convert legacy data to digital form are difficult to make appealing to funders (Lughadha 2004). In discussions about nomenclatural instability in the 1980s and 1990s, the service role came to the fore, and the need for taxonomists directly to address the concerns of "users" was advanced. In the next section I will be discussing in more detail the emergence of the category of user for taxonomy in the context of a database initiative. Those observations largely predate the Rio Earth Summit at which the Convention on Biological Diversity was signed in 1992. Subsequent to this the status of taxonomy as infrastructure, based largely on what I term the marketplace definition, has been established on the global stage. Being treated as infrastructure has hitherto been a dubious honor. While being considered essential gives one a certain amount of leverage, it also means that one risks being taken for granted and neglected in the face of other, more prominent topics. Hoagland argued, in a response from the Association of Systematics Collections to the Convention on Biological Diversity: "In short, taxonomy is largely outside the world economy. It is taken for granted as a free good by governments, resource managers, drug and seed companies, and even by many scientists. People want taxonomy, but not enough to pay for it" (Hoagland 1996: 62). In the contemporary situation, accepting the mantle of infrastructure provider is more welcome than it had been previously, bringing as it does the promise of acceptance as a fundamental necessity and access to new sources of funding. Talk of the taxonomic impediment has offered the occasion to make the case for the need to support and promote systematics.

The formulation of "the taxonomic impediment" to the conservation of the world's biodiversity, laid out in the Convention on Biological Diversity, was influential in establishing systematics as the infrastructure around which meaningful efforts to conserve and understand biodiversity should be organized. Equality of access to taxonomic information was recognized as key to organizing conservation efforts, hence the requirement on signatories to the Convention to share their resources and expertise in this area. The Convention on Biological Diversity has thus helped to reinforce the status of systematics as infrastructure and, by highlighting problems with the current situation, has brought it to the foreground of attention. At the same time, the Convention on Biological Diversity has also been a structuring influence, requiring signatories to organize responses on a national level, but also influencing practices at institutional level by adding additional impetus to the drive toward digital cataloging of specimens.

Although the coming of the Convention on Biological Diversity changed the political and financial landscape of systematics (although for many concerned, not as much as it ought to have done), the advent of the Internet can be said to have changed its aspirations and sense of the possible. As I argued in chapter 3, the Internet has been a potent source of visions for what the discipline could be, and has provided the occasion for specification of both problems the discipline faces and possible solutions to those problems. As chapter 5 described, the Internet could enable what Hilgartner (1995) terms "communication regimes" to be imagined. The Internet could be said to have become a structuring influence on systematics in recent years, seized upon as its model of what it could be even while not determining the specific nature of responses. The structuring influence of the Internet is not straightforward. The Internet, or at least cultural commentaries on it, provides a set of ideas for what could be done and expectations that some of them should be done, but in practice the precise technologies to achieve these visions are rarely readily available. A particular vision has to appear achievable in order to acquire funding, but in order to make a credible application for funding the vision also has to be portrayed as needing some work to achieve. The challenge is to find projects that are plausibly doable but not yet done. Aspirations have to run ahead, but not too far ahead, of capabilities.

The current focus on providing for unified searches across diverse databases has provided just such a challenge. While plausibly doable, it has required considerable effort to make it not just socially feasible (that is, so that institutions and individuals are prepared to cooperate and provide access to data) but also technically feasible, in that functionally equivalent database fields have to be recognized as effectively equivalent in database query language. Much of the technical effort in many of the initiatives described above has revolved around developing ways to store the complexities of biological data, across all kingdoms of organisms and types of data, and working toward standards to allow data to be exchanged between databases, in an environment with many legacy systems in place that would take huge effort to convert to new schemas and new technologies. Standards would provide the infrastructure, which would in turn allow this version of the visions of systematics as universally accessible infrastructure to be realized. The body that has adopted a key role in discussing and establishing standards is TDWG, the Taxonomic Databases Working Group, recently renamed as Biodiversity Information Standards.

TDWG was established to foster cooperation between taxonomic database initiatives and to work toward the development of standards for the representation and sharing of taxonomic data. According to the group's Web site (http://www.tdwg.org/), the first meeting of TDWG in 1985 had a dozen or so participants. From the outset the question of standards for data and the prospect of exchanging data between databases was on the agenda. As expressed in tentative terms at the first meeting in 1985: "The intention was to see what might be done in encouraging institutions and database projects to adopt conventions that would ease the problems of coordination of data exchange" (Taxonomic Databases Working Group 1985). According to the minutes, participants discussed various available candidates for standards and looked at the feasibility of different approaches. In 1989 when I attended the fifth meeting of this group in Las Palmas, Gran Canaria, it was still a relatively small group, conducting informal discussions rather than holding organized presentation sessions. The whole group, at that stage, was able to sit round a large conference table and discuss potential standards and resources. Participants came as representatives of particular initiatives and institutions, but also as committed individuals who were self-consciously exploring the possibilities of information technologies in taxonomy and taking a lead in the field. There was an awareness at that stage that developing databases involved making complex decisions about the nature of taxonomic data, and that many of those creating databases were working on similar problems in deciding on data structures. TDWG provided a place for people to come together, hoping to gain something in terms of a shared expertise. TDWG provided a way of pooling resources, and a certain security to those involved in experimental projects that they would not have to solve all of the problems of taxonomy on their own.

The Web site for the 2004 meeting showed how the organization developed—some 122 registered delegates submitted 52 formal presentations,

9 posters, and 8 demonstrations, and represented major systematics institutions and initiatives. There was a formal constitution and procedure for adopting standards, and the group was endorsed by GBIF as its standards-setting body. The Web site listed 14 adopted standards, with working groups on biological collections data, descriptive data, taxonomic names, and TDWG process, among others. In 2005 substantial funding was made available by the Gordon and Betty Moore Foundation to "modernize" the process by which standards were arrived at, making the process itself more standardized and recommending the use of online tools to encourage participation and coordination (International Working Group on Taxonomic Databases 2005). More recently, TDWG has been working with GBIF on the development of approaches to Globally Unique Identifiers for Biodiversity Informatics (GUIDs), which are intended to provide a stable means of identifying and exchanging data on specimens, biological collections, taxon concepts, and taxonomic publications across a diverse universe of resources on the Web. The process of arriving at a standard involves a complex set of activities evaluating alternative approaches, consulting and recruiting stakeholders and envisaging infrastructures. At this level of granularity the process of standards development once again involves envisaging, orienting to, and at the same time enacting the territory of institutions and initiatives.

The changing nature and status of TDWG attests to the development of the field and the increasing centrality of database initiatives to the work of systematics. Within TDWG, present from the outset but still visible in the standards setting process, is a consciousness of preliminarity. People involved in database initiatives often see themselves as engaged in a project that has a limited life span: although some initiatives, in some circumstances, might portray themselves as complete solutions to problems of biodiversity information availability, most participants have seen too many initiatives and technologies come and go to believe wholeheartedly in such claims. The early TDWG members saw themselves as pioneers, and developing standards was seen as a way of leaving a legacy behind, extending the remit of isolated initiatives in time and space by enabling data sharing between existing projects and with anticipated future projects. Standards were seen also as a way of building prospects for future collaborations and data exchange. The formation of TDWG and the early discussions at meetings reveal an orientation to projects not just as ends in themselves, but as exemplars. TDWG members had a reflexive orientation not only to the goals of a project, but to their qualities as projects and the environment they inhabited in terms of their current positioning and their future role.

It is with standards that it becomes most clear that projects in this field can have a high level of reflexive consciousness that orients itself to something other than the achievement of a particular practical goal. As with other initiatives, I tried to be sensitive to what TDWG meant to people that I interviewed. There was a clear sense that TDWG was an important peer group for initiatives: it was a place to be seen, to showcase one's product and to be kept abreast of what was going on, even where there was some reluctance to be thought subject to decisions that might be taken by the group. Standards have become an important means for this field to codify and share knowledge and to organize resources. As GBIF has grown in significance for the field, TDWG also has acquired an enhanced stature. Initiatives wishing to participate in GBIF and draw on the funding possibilities that GBIF participation brings have perforce to maintain awareness of TDWG processes, and have an interest in shaping the developing standards in ways that suit their needs. TDWG ties together the field of systematic cyberscience by providing venues for mutual awareness between projects and mechanisms for knowledge codification and resource sharing, and by acting as a gateway to accessing funding and participating in influential collaborations.

For a final attempt to depict the complexity of institutional form and initiative interaction that characterizes current efforts in the cyberscience of systematics, it is worth focusing on just one element in the related sites network for the TDWG shown in figure 6.8. The node labeled "CODATA Working Group on Biological Collection Data Access" denotes a working group constituted to develop a specification for information on biological collections, to be expressed as an XML schema. CODATA (Committee on Data for Science and Technology) is a Scientific Committee of the International Council for Science (ICSU), and an example of what Crawford, Shinn, and Sörlin (1992) call the international bureaucracy of science. The working group is formally described as a joint CODATA and TDWG initiative supported by GBIF and BioCASE. It includes representatives from initiatives that themselves offer portals to multiple data providers, including the Australian Virtual Herbarium, the European Natural History Specimen Information Network (ENHSIN), ITIS, Red Mundial de Informacion sobre Biodiversidad (REMIB), and the Species Analyst (TSA). The group is working to develop its own standard and also working with existing standards stemming from the BioCISE project and another standard already adopted by TDWG. In addition to unified strategic discussions, the group has also split into subgroups led by different specialists, focusing on particular areas of content or on specific functional aspects of the standard. This working group therefore weaves together multiple initiatives and meta-initiatives, institutions, and international bodies; organizes its work in multiple ways through a combination of individual effort, committee-based discussion, and small group negotiation; deals with a legacy of preexisting initiatives and standards; and answers to a variety of funding sources and notions of accountability.

Within this section I have discussed not a singular "infrastructure," but varying levels and forms of structuring influence which cross-cut and interactively define one another. The upshot of the various structuring influences provided by standards, technologies, global politics, funding bodies, accountabilities, and institutional histories is a highly dynamic and self-aware field of action. Systematists engaged at this level are caught in a constant dance of getting the right people involved in the right initiatives, spotting the next technologies (and hoping to invest in the ones that other influential initiatives are also committing to), maintaining appropriate branding, and getting the right sets of users and funding bodies involved. Just like the TouchGraph representations that I played with when exploring the field, when a new entity comes into being all the rest shift around to accommodate to it. The fluid TouchGraph representation has become my metaphor for thinking about the field of systematics as cyberscience. Each initiative I have come across is structured in a slightly different way, brings in a slightly different set of players, and envisions its goals in a slightly different manner. This is, though, a thoroughly reflexive field: there is a lot of awareness of who the players are, and constant assessment of what the likely success of the different possibilities is. While the field is made up of a dance of initiatives, this is a very serious dance, since all involved are aware of the stakes in taking a wrong step, in terms of both planetary biodiversity and, more prosaically, institutional survival.

All of the initiatives I have come across have to deal with histories, both their own and those of other projects, and both within their own institutions and interinstitutionally. There is a strong influence from legacy solutions, which have shaped both the way the field stands now and the emotions that participants feel toward it, in terms of what qualities initiatives need to succeed, what the prospects are for change, and what institutions are likely to find appropriate and acceptable. The richness and diversity of initiatives is testament to the availability of funding and to the persuasive visions that characterize ICTs in contemporary society. As we saw in chapter 3, the potential of ICTs as sources of defined product appropriate for funding initiatives has been influential in organizing effort in the field, and productive of creative visions of what the discipline could be. The resulting level of activity is not without problems, since considerable effort is subsequently needed to craft diverse distributed solutions into cumulative achievements. In addition, it can be difficult to acquire the necessary funding to carry out the basic digitization work that is needed as the foundation for other initiatives, but which is less readily represented as visionary.

If I left the subject of the initiative dance without talking about some of the tensions, dilemmas, and compromises that taking part can entail, I would not be doing the field justice. It became clear that not all initiatives were embraced with undivided enthusiasm, and that being a partner in or member of an initiative could take a variety of guises. There were forms of engagement that were wholehearted; and others that saw participation as duties, as necessary evils, or as strategic moves to keep a foothold in a potentially important network. Branding of initiatives was tailored to particular funding outlets and proposals carefully crafted to meet the perceived requirements of different schemes. Institutional interests were carefully balanced against the need to be seen to support meta-initiatives. Interinstitutional collaborations could be the basis for development of high-profile resources, but these collaborations could also entail at the very least close attention from each side as to whether other participants were fulfilling their commitments. It is important to remember when looking at initiatives and the sets of participants and members that they enroll that not all memberships are equal and not all participations unequivocal. Just because an initiative has a set of contributors does not mean that they all take part in the same way, or that they all take the same meanings from their participation.

Each project, and each institution, unfolds a new ethnographic field site that, if explored in detail, would reveal its own values, tensions, and compromises, and its own sets of structures and beliefs, which could be seen as a part of the wider dynamic but also as specific to its own unique culture, history, and structuring influences. This kind of inquiry would reveal participants as, under some circumstances, highly strategic political players and, in other lights, vocationally driven and dedicated individuals focused on furthering systematic insight for its intrinsic interest and pursuing whatever paths are necessary to take them toward that goal. Many participants are also dedicated to conservation issues, focused on solving problems of phylogeny and classification for their key environmental importance. All of these are fair representations. The salient contexts for understanding beliefs and actions vary, as systematists act as participants in the wider disciplinary culture of systematists or biologists, as specialists in a particular group of organisms, and as actors with more locally specific concerns. Recent initiatives in systematics as cyberscience have moved systematists onto the global stage and brought them together in unprecedented ways, but other frames of relevance remain highly significant in orienting responses to the global arena.

The Practical Politics of Cyberscience

In the previous section I looked at initiatives in the application of ICTs to contemporary systematics largely as they relate to one another and to institutions. This outward-facing aspect of initiatives is indeed a highly significant part of their existence, since it is only by forming appropriate relationships with institutions, with other initiatives, with funding bodies, and with users that initiatives ensure their viability as both practical achievements and evocative objects with the right kinds of qualities. If we look only at this aspect, however, we miss out on the local dynamics through which a project is brought into being and sustained as a coherent and credible initiative to those participating in it. Viewed from this more intimate perspective, cyberscience projects are engaged in a practical politics that involves examining and drawing boundaries, making explicit and renegotiating identities, and specifying forms of work. The cyberscience of contemporary systematics, when looked at in this regard, is a thoroughly reflexive activity since it involves a repeated examination of self-identity and of the nature of the environment in which work is done. Successful cyberscience is about making politically potent objects that will be visible to the right people, participate in the right networks, and maintain key institutional identities; it is also about conducting a practical politics that finds ways for diverse sets of people to identify themselves with and invest themselves in concerted efforts.

In this section I focus on one initiative in particular to highlight some of the processes that can go on within a cyberscience project. The project that I look at, ILDIS, was the initiative of a group of scientists concerned with the study of the Leguminosae. This family of plants encompasses almost one-twelfth of the world's flora and is of major economic importance, since it includes many crop plants. The project that I describe aimed to produce an information system to record various data on these plants and make it available to applied biologists. The project participants included representatives from research groups focusing on various aspects and subgroups of the plants concerned, including representatives from the major systematics institutions. The initial goals of the project focused on developing a taxonomic core of data to which other kinds of information could subsequently be attached. The core would contain a taxonomic "backbone" containing a consensus classification, with each of the accepted names cross-referenced to its synonyms.

In this kind of abbreviated account it is easy to stray into a mode of telling that makes the progress of a project seem smooth, inevitable, and without contradictions. This would certainly not be a fair representation of the project: participants have been involved in persuading funding agencies, potential participants, and users of the system, and it is inevitable that the project will have been represented somewhat differently to these diverse audiences. Also, the goals and constitution of the project changed over time as different funding sources became available, as technologies changed, and lately as the global political context for systematics changed. In the current account I necessarily gloss over much of this complexity, for it does not form the core of the points I wish to make. Nonetheless, it is important to remember that this is a distillation.

Probably the most innovative quality of the project was its aim to produce a worldwide consensus classification, a goal not previously attempted on such a scale. The main product was often presented not as a database, but as a classification system that would, as an added bonus, be made available in electronic form. Published classifications of any group of organisms can usually cover only a limited part of the world or a small group of organisms, since they are limited to the scope an individual taxonomist can readily grasp. Even so, authoritative taxonomic treatments are often the work of a career. There was, prior to the ILDIS project, no mechanism for this patchwork of often conflicting taxonomic treatments to be stitched together into a coherent whole. The legitimacy of the ultimate product would depend on recruiting the appropriate specialists and institutions to give adequate coverage and to be seen to have the blessing of acknowledged authorities. The approach that was taken in the ILDIS project was to draw up initial checklists on a geographic basis, then call on taxonomic coordinators to focus on a particular grouping and reconcile conflicts arising from the merger of geographic lists. These taxonomic coordinators were required to make reasoned assessments of the best consensus position, based on their own knowledge and consultation of other experts. The qualities of an appropriate coordinator were described in a grant proposal thus: "Where several experts are known, care has been taken either to contact one of known moderation of views, or, where several experts have regional knowledge of widespread taxa, the experts are being asked to collaborate" (Grant proposal case for support, 14 September 1990). The scientific consensus was therefore to be achieved via some carefully thought out social mechanisms, designed to maximize the input of appropriate expertise and the likelihood of achieving a broadly acceptable end result.

The end product of the coordination process was intended to be a stable taxonomic backbone, offering a standardized classification scheme and nomenclature to which other information could then be attached, either within the project or by additional user groups, such as applied botanists or conservation groups. I have previously described the ILDIS project as involving "disappearing plants and invisible networks" (Hine 1995). The information system, in providing information on plants in an accessible and organized fashion, was to stand in for plants themselves. This much is typical of the standard "cascade of representations" (Latour 1987) that characterizes science, as successive transformations turn observed phenomena into increasingly abstracted representations. The information system mediates for plants, providing instead a set of information about plants that, it is supposed, users will be able to access and link up with the plants they encounter in whatever field of application they work.

The information system is also supposed to turn the taxonomists who produced it into an invisible network as far as users are concerned. One of its aims was to provide a way to manage the tensions between the scientific ethos of taxonomy and its responsibilities to users. By situating the information system to users as a provider of taxonomic information, and to taxonomists as a user of taxonomic information, it would stand in between the two groups and allow for autonomy. Taxonomists would not usually engage in collaborative production of a consensus classification. In this instance, taxonomists were persuaded to participate in the production of consensus by the promise that this would meet the needs of users for stability, and yet still allow them to work unhindered in their usual ways. As one grant proposal in connection with the project put it: "This coordination and consensus-seeking will in no way seek to suppress the regular work of taxonomists in proposing new classifications and updating treatments to take account of new discoveries. We are simply trying to decouple the necessary flux and debate that goes on amongst taxonomists from the single consensus system that might responsibly be made available to the rest of biology as a stable reference system" (Grant proposal case for support, 14 September 1990). The information system would thus, it was claimed, largely protect users from the vagaries of changing taxonomic opinion and by taxonomically intelligent searches enable the correct data to be retrieved whether synonyms or accepted names were being used. By sitting on the boundary and representing each group to the other, the information system was intended to resolve tensions between them.

In describing the process as one of sitting on a boundary, I risk giving the impression that the groups on either side preexisted in a stable form. This is not the case for the scientific ethos and processes of taxonomy. Early stages of the project involved specifying precisely what it was that taxonomists did and defining how conflicts might be expected to arise and be resolved. The same proposal I quoted above described how the activities of coordinators were "similar to the editorial work undertaken by the writers of Flora," making explicit links to existing practices. The causes of conflicting taxonomies were specified as twofold: "In our view the number of *real* cases of conflict is small. This is when two taxonomic experts presented with the same plants, same data and same techniques persist in proposing unreconcilable systems. More frequently the experts formed different opinions because they have seen different subsets of data—either they are in different continents, have limited access to materials, or have access only to different data sources (morphology, pollen, breeding, phytochemistry etc.)" (Grant proposal case for support, 14 September 1990; emphasis in original). This kind of detailed specification of taxonomic work does not preexist the project, but is produced in response to the particular situation entailed in writing a grant proposal to a science research council to fund a consensus-seeking initiative.

In similar style, it could be argued that the users, for whom the information system was to provide a stable nomenclature, did not preexist the project. At various stages different means of offering services and products were explored, including licensing the database, selling subscriptions to paper-based or electronic versions, providing tailored information services, and forming alliances with third-party service providers who might repackage the data. All of these discussions raised the specter of user needs and practices, but it was never clear even whom might be consulted about their preferences. The group of users who were to be protected from the nuances of scientific taxonomy were key in much of the project's deliberations and its publicity. It remained, however, a troubling concern to identify and enroll user representatives and to gain a clearer picture of how they might use the system. The category of user in relation to taxonomy is one that has been latent in the field's descriptions of itself as providing the infrastructure for the rest of biology, but has rarely been specifically defined or claimed by a particular set of people. In relation to information systems such as the one described here, the specification of users has acquired a new significance, not least because access to some kinds of funding depends on making a credible case about user needs. In this regard there is a contrast with the collections databases described in chapter 4, where it was more acceptable to make broad assumptions about use in the context of an overall archival mission.

The project thus enacted a boundary that defined the identities of the groups on either side, as a part of an overall claim to resolve tensions within taxonomy between carrying out scientific activities and providing useful services. It is not, however, altogether fair to represent the system as resolving the science–utility tension, at least insofar as relationships with funding bodies are concerned. One grant proposal was turned down by a science research council on the basis that it did not contain sufficient scientific content: "The opinion of the Subcommittee is that the application proposed an exercise in information collation and did not contain the essential elements of scientific research required by the Subcommittee. The proposal did not involve any new experimentation or observations, but rather its approach was clerical, encouraging consensus decision by increasing taxonomists' awareness of already available information" (letter to ILDIS, 9 January 1991). The main organizer of the ILDIS project subsequently defended the scientific status of the work involved in achieving consensus in a resubmission of the proposal: "It should be stressed that the 'consensus' referred to is not a social event achieved say by voting, but a scientific event achieved by meticulous analysis of comparative data ensuring that all of the data available to each of them is incorporated into the synthesis. It is at this point that experts who have previously disagreed see that some of the distinctions which they had previously sustained may no longer be valid against the background of the larger data set" (letter to funding agency, 15 November 1991). The form of this disagreement is quite specific to the funding agency concerned: in applications to charitable foundations, the economic and humanitarian importance of the particular group of plants involved was stressed. Nonetheless, one of the key ways in which the project was presented to funding agencies was as a mediator between taxonomists and users, even though this solution to the utility–science tension was not always one that was acceptable to scientific funding bodies.

The identity of the project itself was a carefully crafted achievement. As an international collaborative organization it was different in a number of respects from usual ways of conducting scientific work, not least in that it had legal status as a registered charity in the UK, and that it formed an executive, a board of trustees to act as the official owner of the database, and a range of special purpose committees. Formal agreements were entered into with various institutions to agree the form of collaboration and the ownership of any data contributed. A joint publication agreement was signed with a commercial publisher, and funding acquired from institutions, from research councils, from charitable foundations, and through royalties. Through relationships with these bodies, the project aimed to present an identity that provided a stable basis for collaboration and conducting business. Acquiring legal status was seen as a means to entering into formal relationships as more than the sum of the individual taxonomists and institutions involved in the project. More prosaically, the project acquired a corporate identity by adopting a logo that featured on a range of printed leaflets and letter heads and subsequently on the Web site. Discussions about the choice of logo expressed some concerns about the image that was portrayed. Some concerns focused on the loss of identifiable images of plants from the logo, and others revolved around whether the new image was appropriately smart and professional. The file containing correspondence on this issue included a number of leaflets from other projects and institutions, included to allow for comparison with the kind of image that was being portrayed elsewhere.

In presenting itself to outsiders, the project had always to attend to the motivations for taxonomists and their employing institutions to participate. Most obviously this related to concerns that credit needed to be given to those participating, so that users would know whose data was being used. The nature of the consensus-seeking process meant that many people might have been involved in providing and ratifying any particular piece of data. A list of credits was therefore displayed on starting up the database, and lists of coordinating centers and contributing institutions included in project publicity. Credit and motivation in a variety of guises formed a recurring topic for the board and executive. This practical politics was supplemented by a sensitivity also to institutional and international politics: awareness of which institutions saw themselves as competitors, and awareness of international sensitivities that made it difficult, for example, for Russian experts to participate in early stages of the project.

This kind of practical crafting of a project in which it would be possible and desirable for the key parties to participate illustrates the complexity of coordinating the project. The coordination involved such heterogeneous concerns as: choice of participants and motivation of contributors; recruitment of appropriate user representatives; the sequencing of activities including the provision of appropriate software and taxonomic contributions; the constant quest for funding with a diverse set of funding bodies; maintenance of administrative and communicative structures to hold the project together; balancing the current state of technologies with the capacities that could be expected from contributors; and the specification of data entry fields and the quest for data integrity. Coordinating the project involved coordinating technical, social, and scientific concerns in the present day, such that it became a rich and complex model of the discipline itself. It also entailed a conscious orientation toward the past and the future of the discipline.

The orientation toward past and future helped to provide a sense of the project as part of a cumulative development. It was often described as the upshot, or natural progression, from a previous database project focused on a more restricted group of plants. Against this past experience, and based on its current foundations and self-consciously innovative identity, it was also keenly promoted as part of the future of the discipline. When I wrote a previous paper on the ILDIS system, I had initially changed the name of the project in order to protect the confidentiality of those concerned. As I noted in the published version of that paper, the project organizer asked me to reinstate the real name of the project, on the basis that "Given that we are openly trying to promote ILDIS as a prototype in international/scientific organizations, I think ILDIS would be proud to get credit for what you describe!" (Hine 1995: 83). He also wrote publicly at various stages of the potential of the project as a prototype for other species information systems (Bisby 1986, 1993). The project was therefore positioned to be a part of the next wave of developments in the discipline. Although the visions for that next wave have changed over time and a number of larger-scale projects have been discussed and abandoned, ILDIS has indeed become part of the emergent next wave, organized around current moves to provide common portals for access to biodiversity information.

The ILDIS project has formed the seed for a concerted effort to provide nomenclatural databases for all species, through a project called Species 2000. Organized by the same lead individual as ILDIS, this project aims to provide stable taxonomic backbones for biodiversity information through a system of federated database. Species 2000 continues the ILDIS focus on databases as an interface to allow a concerted front to nonspecialist users while allowing for diverse practices among taxonomists to continue. In Species 2000 an added level of complexity is offered by a further portal bringing together diverse species databases under a common interface: this considerable technical challenge is deemed important in order to allow for the use of existing legacy databases while protecting users from the complexity of diverse platforms. The legacy situation, working with existing databases rather than carving up the problem domain in a logical fashion, means that there are inevitable differences in data structure and in taxonomic terms, involving both overlaps between the existing databases and areas where no candidate databases exist. The federated database solution is not unique to systematics, but has been widely adopted in commercial applications as a means of retaining autonomous individual databases while giving users "an illusion of interacting with one single information system" (Ziegler and Dittrich 2004: 3). There are considerable technical challenges on the way to achieving this goal, as Ziegler and Dittrich describe in their review of three decades of attempts to achieve such seamless integration.

The ILDIS project acted as a pioneer also in identifying areas where computer science solutions are either required to achieve working solutions, or can assist in making the task more accurate and efficient. Building the ILDIS database involved merging checklists from different sources, and provided the occasion to develop taxonomically intelligent software solutions to automate the process as far as possible. Crucially for the presentation of this project as a fundable problem, the issue was generalizable to other kinds of taxonomic database than just the ILDIS project (Embury et al. 2001). The technical challenges of combining databases, as proposed by Species 2000, offer another set of problems with a generalizability to other domains where heterogeneous data is combined (Xu et al. 2001). By construing problems as generalizable issues, and by publishing solutions as general purpose contributions, these projects present themselves as prototypes. Species 2000 has continued the strategy pioneered by ILDIS in exploring a diverse array of funding possibilities, presenting some aspects of the project as technically innovative for research council funding, while other aspects tap more directly into funding available through national governments, the European Commission, and GBIF to address availability of biodiversity information.

The federated approach adopted by Species 2000 provides leverage to bring potentially competing databases into some form of cooperation, and also provides leverage to bid for funding to address objectively identified gaps in coverage. Although the question of who exactly users are and what they want is no more resolved for the current federated database initiatives than it was for ILDIS, the concern has in part shifted thanks to the increasing presence of meta-initiatives. In collaboration with ITIS, Species 2000 has been adopted by GBIF as provider of the taxonomic component for the envisioned metaportals of biodiversity information. GBIF therefore figures, for Species 2000 and its component databases, as a kind of "meta-user," legitimating their efforts and providing additional leverage in promoting the importance of the individual projects to funding authorities and data contributors. The success of Species 2000 in enrolling the widest possible community of appropriate participants is noted with some irony in a funding application made by the group, in answer to the question prompting them to name independent referees who could be approached to assess the application: "Species 2000 is developing as such a wide community project that it is difficult to name independent people not already involved. . . . For this reason we provide four names, none wholly independent" (Bisby,

Prance, and Henderson 1998). The case for support makes clear that the project leaders see their own work as both technical and political.

It is at this point that a sociologist of science is tempted to step back, raise the hands, and admit defeat: the level of reflexive awareness and sociological understanding that has gone into making Species 2000 is quite disarming for the sociologist. By explicitly framing problems as biological, technical, and social, and by the complex understanding of practical politics that is displayed, the Species 2000 approach provides little for a sociological analysis to layer on top. The members of the Species 2000 project team appear to exemplify Callon's description of engineer-sociologists (Callon 1987) and become systematist-sociologists, seamlessly but also quite consciously crossing between aspects of their work they view as technical, biological, and political, crafting solutions that meet the needs they recognize in each domain. In systematics as cyberscience we see, then, a kind of science taking shape that involves scientists as reflexively aware rounded beings and projects as multidimensional heterogeneous endeavors.

The picture of systems design as practical politics gained from the ILDIS project was mirrored by other systems designers to whom I spoke. They portrayed themselves as working out which user requests and what political pressures to accede to and which to resist, attending to funding structures and lines of accountability, as well as visions of the discipline and ideas of appropriate directions in which to travel. Users figured as the rationale for projects and as their problematic, both in recruiting sufficient numbers or sufficiently impressive representatives, and in working out what exactly these users might want from the system. Working practices were specified and crafted as projects developed. Software developers and advisory groups acted as gatekeepers for disciplinary development, pushing projects in directions that they found socially, financially, technically, and scientifically feasible. A microperspective on initiatives therefore shows that their public presentations are finely tuned and carefully crafted, and that their achievement as practical solutions is by no means guaranteed by the availability of the technology alone. Technologies, and the broader cultural expectations that surround them, provide some of the visions for directions in which to push the discipline, but the realization of these visions is a matter for work across a broad range of fronts.

Individuals and the Recognition of Achievement

In discussing initiatives thus far I have tended to focus on the concerns of institutions, looking at the attention to funding structures, policy developments, and relations with other institutions that informs decisions to take part in initiatives. It is important, though, to remember that institutions are made up of individuals, and that taking part has to be a meaningful activity at the level of the individual as well. In some cases individuals will be nominated by their institutions to act as representatives in regard of particular initiatives. In other cases, a database project will grow from a particular individual's interests, often as the upshot of a research project that accumulates data in electronic format. In these cases, developing a database and making it available via the World Wide Web can be a way of extending the reach of a project, making it visible in a different format to the printed monograph and potentially enrolling new audiences. It is certainly possible to accrue reputation in this way. Professor Humphries, of the Systematics Association, used one such resource as an exemplar in his evidence to the House of Lords Select Committee: "Paul Williams' own website dealing with bumble bees, of which there are 240-odd species, has taken him most of his career to work up into a reasonably satisfactory knowledge base, and his website is terrific—as a result of him knowing the organisms well and having a talent for writing those kinds of things" (Select Committee on Science and Technology 2002b: 132). In the discussion of the practical politics of cyberscience initiatives in systematics, I referred to the vision promoted by the key organizer, who plays a significant role in shaping the initial project, recruiting other participants to the cause, and creating the conditions for the successor project to emerge. Many participants attribute the successes of the project to the leader's charismatic persuasiveness. As the main promoter of the ILDIS project and lately a key instigator of Species 2000, Bisby has been able to forge a distinctive career for himself as a specialist in this type of work, listing his research interests on his departmental Web page as "Species diversity & taxonomic information systems; The Internet & GRID as a global biodiversity information network. Taxonomically intelligent systems: Legume Taxonomy—gorses & brooms (Genisteae) and vetches & peas (Vicieae)." Notably, and unusually for a professor in plant sciences, the specification of the plant groups in which he is expert are stated last, after the information systems interests for which he is now probably more widely known. It is clearly possible to make a credible career as a cybersystematist, although it is by no means common.

Individual contributions to initiatives can take many forms, including rousing support and team building as well as making heroic individual efforts. Less prominent forms of involvement should not be forgotten: each project requires some kind of technical or programming support, and many require considerable effort in data entry, checking, and editing. In addition, projects aiming at authoritative status require considerable input of taxonomic expertise to verify information and adjudicate over conflicting data. All of these contributions need to be meaningful in some sense to the individuals as well as to the institutions. In the remainder of this section I will therefore examine the existing reputational system of systematics, and consider the extent to which the effort of individuals within initiatives is visible and open to rewards.

The contribution of individual systematists to the field is notoriously difficult to pin down using the standard techniques of citation analysis (Krell 2000, 2002; Valdecasas, Castroviejo, and Marcus 2000; van der Velde 2001). Small numbers of experts tend to be working in any given specialism at a time, and major works are published infrequently, so there is room for assessments of contributions within the community to be made on the basis of reputations: as we saw in the choice of taxonomic coordinators for the ILDIS project, it was possible to select candidates of "known moderation." I am tempted to suggest that this kind of reputational system for taxonomic contributions is mirrored in cybersystematics, where some people have acquired a reputation for their work and are respected accordingly. In systematics institutions, reputations for taxonomic work can lead to recognition within the institution, and this can also happen where a reputation is acquired for significant contribution to taxonomic information systems. Universities are often said to be put off by the lack of formal citation recognition that systematists gain and the low levels of external grant funding that they bring in, leading to reluctance to hire them in relation to other biological specialists and a resulting overall decline in the number of taxonomists at universities. The ILDIS project and its promoter were unusual in this regard, being based within a university context even though there were few mainstream journal articles directly attributable to the project. Increased funding opportunities for cyberscience may make systematics attractive again to some universities, since there is a possibility for staff engaged in this kind of work to bring in external funding; but as yet the ILDIS project and its leader remain exceptional.

Much more numerous than the individuals who take a prominent role in promoting projects and visions are the programmers who realize projects, and the taxonomists who contribute data. In the early days of the ILDIS project there were few information technology specialists in taxonomy. Three taxonomists took on the majority of the work of programming the database and providing software for merging and checking data. They had some success in promoting the software as a general-purpose tool for systematics, and thus were able to transform themselves into visible specialists instead of having their work submerged into the overall project. There were, however, some concerns among others on the project that it was becoming overly reliant on the work of these few individuals. There is often pressure to move away from a reliance on individuals toward a more "professional" model in which software development is viewed as owned by the project. The BRAHMS software for specimen collection management has been promoted as a general-purpose tool, and though its author has invested considerable efforts in developing and supporting the software it has also been important, through the establishment of an advisory committee and the presentation of a depersonalized brand identity, not to let the project be seen as a sole endeavor.

The developers of software for taxonomic databases thus seem to face some dilemmas, needing to build careers and make their contributions institutionally visible, while also feeling the need to present a depersonalized face for projects, lest potential users be put off by reliance on individuals. Adopting a database solution is a considerable investment, and users want to be reassured that there is a broad base of support for the one they adopt. There appears to be a difference here between the work of programming information systems and that of producing another form of software well-established in systematics, for phylogenetic analysis. In information systems work, the authors struggle to manage tensions between gaining individual credit and publicly presenting a depersonalized face for the generalized solution they hope to persuade others to adopt. By contrast, phylogenetic analysis software is often portrayed as a much more individualized product. It has certainly been possible for authors of phylogenetic analysis software to develop high-profile individual reputations. A key example is David Swofford, author of PAUP. His Web site lists his research interests as: "The theory and methodology of phylogenetic inference from molecular sequences and morphological data (development of algorithms and software to facilitate model-based analysis of DNA sequences; comparison of optimality criteria used to evaluate evolutionary trees; methods of combining molecular and morphological data using maximum likelihood; assessing robustness of maximum-likelihood methods to model violation; exploration and evaluation of metaheuristics useful in searching for optimal evolutionary trees)." This presentation clearly positions software development as a means to realize a theoretical and methodological contribution.

It is possible that the taboos against using a system developed by an individual do not figure so strongly for phylogenetic analysis as they do for information systems. Users do make some investment in learning to use an analysis package, but the burden does not compare in any way with the investment that is made in choosing a system for specimen data. Any reluctance to use work that relies on a single individual is offset by the scientific urge to use algorithms from documented sources, and to know which work to cite to give a foundation for analysis depending on a particular program.

Despite the in-depth appreciation of the niceties of taxonomic data required to write an appropriate information system, the work itself is rarely seen as a scientific contribution. In contrast, phylogenetic analysis software is seen not only as a tool, but as a manifestation of a piece of theoretical work, giving it a scientific respectability not automatically shared by the authors of less obviously analytic software.

Phylogenetic software is routinely cited in the standard scientific style, although the usual author-date reference often corresponds to the software itself, or to the user manual. Authors of successful phylogenetic software are able, through this practice, to acquire citation profiles far out of the reach of taxonomists specializing in the description and classification of organisms or those systematists focusing on information systems. Computer programs for phylogenetic analysis are thoroughly rooted in a rich tradition of academic literature debating theoretical import and methodological choices, although it is unlikely that this literature will be cited in any depth in a paper making use of the program for practical purposes. This is a heterogeneous literature, in which programs point to theoretical debates in the literature and vice versa. Algorithms are published and peer reviewed in journals, and it is these articles that are cited in theoretical discussions, while the programs are cited in articles that use them to make phylogenetic judgments with only limited citations of the theoretical literature. Students are encouraged in textbooks to increase their understanding of phylogenetic analysis by gaining experience of appropriate computer packages (Schuh 2000). Computing therefore counts as both theoretical work and practical experience in systematics. Whether the program is presented as an unexamined "black box" or not depends on the particular venue in which use is being presented and the audience for which that work is being done. There is an active peer community in the development of algorithms for phylogenetic analysis for whom black-boxing is a highly inappropriate term. In the literature making practical use of analysis packages, the traditional citation format keeps open a pathway into the "un-black-boxed" realm even though it does not discuss it in any detail. This literature depends on the standard model of contributions attaching to, and accruing credit to, individual scientists.

This is not, however, to say that scientific reputations or appreciation of the qualities of the algorithm alone lead users to prefer a particular solution. Reviews of available software packages often focus on such issues as flexibility, hardware requirements, and speed. For example, Schuh begins his appendix reviewing selection and acquisition of software thus:

> Computer assisted phylogenetic analysis can consist of as many as four distinct activities. These are:

- Data matrix preparation
- Nucleotide sequence alignment
- Phylogenetic computation
- Cladogram analysis and printing

Success in performing these activities will depend on your ability to acquire and master suitable software. The factors likely to affect your choice among available options will include: operating system choice, speed, ease of use, and price. (Schuh 2000)

Availability and usability are thus clearly thought to be relevant factors, along with any algorithmic preferences, in determining which package is used. Indeed, there has been some suggestion that these factors might have affected the development of systematics. Hull (1988) suggests that part of the success of cladistics over phenetics was encouraged by the contrast in complexity of analysis and level of expertise. Pheneticists were expected to become connoisseurs of algorithms, trying out different numerical methods on data. In contrast, "The cladists presented them with a method—one method—and they could use it without becoming experts" (Hull 1988: 520). Of course, this says a lot about the way approaches were packaged and the sets of expectations that revolved around the different approaches rather than necessarily mapping an absolute distinction between them.

To return to the key point of this section, however: whatever the reasons for choice between different phylogenetic analysis packages, it is clear that scientific credit can and does accrue to their designers for the work invested in them. This is also becoming a crowded field of many competitors. Felsenstein (n.d.) cites some 227 packages that he knows of, dramatically illustrating the richness of the field by the collection of logos with which he decorates his site (http://evolution.genetics.washington.edu/phylip/software.htm). Although not all will develop major reputations, it is clear that many see the effort of developing and distributing a program as a worthwhile investment of effort. By contrast, scientific credit is less readily available for the individuals who carry out technical work on systematic databases and contemporary distributed information systems. Their career advancement depends on institutional recognition for their efforts, more likely to revolve around their crucial role in making funded initiatives feasible.

If work in systematic databases tends to be less visible in terms of conventional scientific credit as manifested in the formal published literature, we can ask whether it is being substituted by alternative forms of credit specific to the medium. Publishing their own data in electronic form and making it available through their own Web sites means that initiatives could make their own decisions about the extent to which the various forms of work they entail merit public recognition. Whom, then, do initiatives give credit to, and how do they do it? A straightforward means to address this question is to explore project Web sites and databases, noting the places in which individuals are mentioned in connection with the project.

The ILDIS project did discuss at some length the extent to which recognition could be given for the work of the taxonomic coordinators who drew up checklists and ensured their consistency. In many cases the contribution that coordinators made either was an extension of existing projects, or led to publication of checklists in traditional printed format as well. Nonetheless, a lot of effort was asked of contributors, inevitably diverting them from other forms of work and asking for investment both from their institutions and from them personally. The ILDIS Web site gives prominent place to lists of taxonomic coordinators, of trustees, and of regional centers. Querying the database for a given taxon returns a record with the source of the data and the name of the taxonomic coordinator who checked it, serving the dual purpose of lending traceability to the data and giving credit to the individuals concerned. As well as the taxonomic input from coordinators, other kinds of project contributions were also publicly noted. Both the ILDIS site and Species 2000 give the name of the Web site designer. Species 2000 posts a project team of ten people, and a list of members that consists both of individuals and databases "represented by" individuals. A technical team of twelve people is credited as giving Web site and software support. Documents within the site detailing data standards and connection protocols are authored by named individuals.

Species 2000 uses a federated databases model. A query of the combined annual checklist returns results marked with the logo of the contributing database. Each record is displayed with a "Latest taxonomic scrutiny" field, which can contain the name of an individual and a date, although this information is not available for all records. In most cases it is possible to click through to the originating database. The kind of information presented then varies between databases. FishBase returns the taxonomic coordinator for each record, which in turn can be clicked on for a page of information about that person's contribution. AlgaeBase gives the names of the creator of the record and an updater of the record.

Some databases, particularly those such as ITIS which is an amalgamation of other databases, are able to give little detail about the efforts that have gone into their creation. FishBase is unusual in keeping detailed records, including a page on the Web site of pictures of the team, past and present, each linked to a page giving information about their contribution including activities and dates. LepIndex gives its authorship by a list of four names on the front page of the Web site, plus the name of a Web site designer, but a further page describing "project personnel and their roles" lists some sixteen individuals including those responsible for programming, data entry, and data checking. Each record in the database may have a field recording those who approved it and last updated it. These two databases are unusual for the extent to which the various forms of work that have gone into them are recorded. At the opposite extreme, BioCASE, as a large European project, is populated largely by institutions, although a text on the Web site also comments on the expertise of certain key individuals within these institutions. It is possible from the Web site to work out who the coordinator is and who are some key individuals, especially those on the technical team, but there is little other trace of individual effort, and people tend to participate as institutional representatives.

In this section I have examined the relationship of individuals to cyberscience initiatives in systematics, asking how far the work of systematists as individuals is observable within the networks of initiatives and institutions, and examining how far credit accrues to individuals for their involvement. The leader of the ILDIS and Species 2000 projects is not alone in taking on a role as promoter of ICTs in systematics. Such individuals are often not particularly visible on the Web sites and public presentations of initiatives, taking care as good practical politicians to present the full spectrum of participants. Nonetheless, many of those involved in the various initiatives and collaborative programs are aware of key individuals whom they recognize as leaders. Within initiatives and institutions, then, are motivated individuals who shape developments and persuade others to join them. As practical politicians they make it possible for others to participate, and they pursue the connections to make projects achievable. However, these individuals are acknowledged by all to be rare. Much of the more mundane labor of data entry, editing, and checking tends to disappear from the public record, even though some projects like FishBase and LepIndex do make efforts to record input of this kind. More of the traces of coordinating work and programming work remain, as do records of the authorship of documents and Web sites. The key concern that projects have is making sure that taxonomic judgments are duly credited, since this both ensures that the pedigree of data is traceable and offers a potential source of credit for the taxonomist concerned.

Many systematists would say that they are motivated as much by their engagement with the task of understanding the particular group of organisms that interests them as they are by any abstract urge to gain scientific credit for their work. This kind of engagement with the organisms is the motivation that makes sense of many contributions to systematic databases, which may be seen as a seamless extension of existing work rather than a separate kind of cyberactivity. This is the motivation which the invitation from IPNI (International Plant Names Index 2004) to its users to report any errors that they come across, or indeed adopt a group of organisms for close data scrutiny, invokes. The hope is that any systematist who came across data he or she knew was wrong would almost be unable to resist doing something to correct it. Many individuals will, however, need to make sure that their participation in any project or activity that takes substantial time is meaningful in an institutional sense, either by acting as representatives of their institution or by initiating a project that benefits the institution, through attracting funding, maintaining the institution's presence in a high-profile initiative, or adding to the institution's existing portfolio. Career advancement, or at least career security, may depend on being involved in institutionally significant work as much as accruing scientific credit in a more traditional sense or satisfying an inherent love of the field. This is particularly true in the current institutional landscape, now that so few systematists work within university departments: the focus on citation measurements may be diminished, but the institutions operate within a territory that feels increasingly politicized and individuals must secure their position by being seen as doing valuable work.

Conclusion

This chapter has been about the complexity of the contemporary landscape of systematics. I have been concerned to demonstrate that the territory of information and communication technology development for systematics comprises an intricate interweaving of individuals, initiatives, and institutions. It appears that Latour and Woolgar's (1986) description of scientists operating within credibility cycles, accruing symbolic and material capital through successive phases of research funding, data generation, publication, and renewed funding, is too simplistic to portray this situation. The credibility cycle for the systematist takes different forms, given the focus on reputation rather than publication and the relatively low importance of external funding until recent years. Institutions and initiatives have their own credibility cycles which coexist with those of individual systematists, thus mutually reinforcing one another. A single cycle would be insufficient to portray this level of convolution, and it is for this reason that I have fallen back on metaphors such as the "dance of initiatives" to portray the field. The territory within which a contemporary systematist operates is multifaceted. As Crawford, Shinn, and Sörlin describe it, scientists are rounded individuals with multiple allegiances: "Scientists, like other people, bore identities, they belonged somewhere, and they were loyal to something. Even more importantly, the daily activities of scientists were carried out in a framework of institutions, agendas, career opportunities, working language, financial support and patronage systems" (1992: 45). For the systematist, cyberscience has brought about new allegiances and altered the landscape, in a way that both awakens visions of the possible and yet prompts appreciation of the security of the traditional institutional structures.

There has been a proliferation of branded initiatives on the Internet in the sphere of systematics, focusing on the need to be recognized for effort invested. In electronic services, then, efforts are made to make clear to visitors the identity of the resource they are visiting through strong branding. Within many initiatives, though, the role of the traditional institutions is still highly significant. As repositories of collections of material objects that require funding to maintain, they have to pay considerable attention to their image and to convincing others of their worth. They have both developed their own branded initiatives and made certain that their contributions to cross-institutional initiatives are visible. Individual reputations are also considered key, as a part of the emphasis on reputation within systematics and the belief that the quality both of information and of taxonomic judgment varies. Products are not generally branded at the level of the individual scientist, although each piece of information will be marked according to its origins so that reputational judgments can continue to be made. To return to Poster's (2004) expectations about the impact of digital commodities on branding, we can see that the effects are not as directly associated with the technology as he proposes. Considerable effort within systematics has gone into developing forms of branding and attribution of credit within digital resources that will be seen as appropriate to the expectations of the discipline. Systematics has, as yet, little experience of the ways in which its resources will be interpreted in practice by users. The expectation being built into current implementations is that the scientist, as an individual, continues to matter as the originator of scientific authority.

Across the entire complex territory of contemporary systematics there is a vast amount of second-guessing. This focuses on the likely technical success of various projects, the changing political and funding priorities, and the moves likely to be made by competitors. One of the key areas of second-guessing is the identity of users. The category of "user" has been of key importance in promoting openly available taxonomic information resources, yet it also remains highly problematic. As discussed for phylogenetic analysis software, it is unclear whether preference for the scientific qualities of a particular algorithm has much influence in the face of concerns about cost, processing power, speed, and usability. Concerns that users are not sufficiently sophisticated to know high-quality taxonomic information when they see it have driven many of the developments within taxonomic information systems. There has been a widespread fear that if taxonomists do not develop such systems then users may turn to other sources for their information, relying on unverified data and leaving the taxonomic community sidelined. There is, though, little certainty who users of these resources will be, how informed and active they will be as consumers, and how receptive they will be to the nuances of source and reputation. As Haas (2006) describes it, taxonomists and users are seen as "two species that rarely meet." Ultimately, then, although the branding of initiatives and the traces of individual and institutional effort that are being embedded in contemporary systems depict current beliefs about what might be important to users, there is no certainty on how they will be received. As a cyberscience in the making systematics has moved beyond the security of established relationships, and it awaits a new audience with a nervous yet hopeful air of anticipation.

7 Conclusion: Socially Meaningful Cyberscience

In trying to find out how systematics has been experiencing its engagement with ICTs in terms of change and continuity, I have ranged across some diverse aspects of the discipline and have found multiple ways in which ICTs are involved in the work of making systematics a practical endeavor for the contemporary world. At the same time, I have been engaged in making my own kinds of sense, finding ways to connect my emerging field with a diverse set of concerns that resonate for science policy and practice and for my (inter)disciplinary homes in science and technology studies, Internet research, and sociology. In this chapter I will reflect on where my connective explorations of systematics as cyberscience have taken me in terms of conclusions for these diverse concerns. These reflections fall into four sections. In the first I focus on the experience of systematics with ICTs, drawing together the stories told in chapters 3, 4, 5 and 6 into a summary of the role ICTs play in sustaining contemporary systematics. In the next section I situate the experience of systematics in the wider context of the relationship between cyberscience, change, and specificity, aiming to explore further how it is that ICTs can act as a resource for imagining without determining outcomes. This then forms the basis for reflecting more pragmatically on the potential of ICTs as a policy instrument for promoting desirable changes in science. I then finish by thinking about the limits and the potential of the connective approach that I have taken in the context of other approaches to the study of science.

ICTs as Sensible Systematic Practice

ICTs have become an increasingly prominent part of the disciplinary reality for systematics in recent years. Everyday practice for the individual taxonomist often entails the use of email as a routine form of informal communication and the habitual reference to electronic literature sources.

Systematists often rely on software packages for performing phylogenetic analysis, and use both personal databases for storing data on projects and institutional databases, locally and further afield, for tracking down specimens and literature of interest. For some individuals the development of information resources has become an activity in its own right, whether working within an institutionally sanctioned initiative or acting as an institutional representative on a larger project. Involvement might constitute provision of data or taxonomic expertise in checking the integrity of data, or might focus on developing technical aspects of software packages, Web sites, and data standards. Efforts have been made by database designers to ensure that traces of the source of intellectual input are left behind so that taking part in these projects will count as a meaningful thing to do in terms of career development and reputation. Some technical work is still undertaken on a craft basis, by taxonomists who happen to be particularly technically inclined; but increasingly there is a cadre of individuals focused on bioinformatic aspects of taxonomy, for whom this is a career path in its own right.

For institutions, ICTs are a significant part of the complex political landscape and funding climate that they must inhabit. Acknowledging the common perception of ICTs as representative of efficiency and progress and aware of the policy steer toward their use, institutions have deployed these technologies as a means to reinforce their position as providers of fundamental information for the conservation of biodiversity and as representatives of an infrastructural science which governments neglect at their peril. ICTs have been appropriated in the cause of saving the discipline from neglect and underfunding, and as a means to underpin the viability of the major institutions. However, the current climate dictates that although institutions have to be aware of the need to protect their intellectual property, they also acknowledge that initiatives based around single institutions may seem insufficient to meet expectations of seamless access to information and grand visions organized around new technologies. New organizational structures have come into being, representing meta-initiatives that rise to the challenge posed by expectations surrounding ICTs and in turn require systematists to think about the consolidation of knowledge on an unprecedented scale. Standards-setting bodies have evolved a new significance in their role of mediation between the rich diversity of legacy initiatives and the grand dreams of seamless access. The current picture of systematics as cyberscience thus both reflects the prior institutional landscape and spins a new web of initiatives across it, tying institutions together in new and more intricate ways.

Systematics is a cyberscience in that it has become suffused with ICTs and ICTs have become a potent way to imagine what the discipline should be and to understand its past and present. Projects involving use of the Internet to provide access to information have become an increasingly significant part of the way the discipline presents itself to outsiders, including funding bodies, governments, and potential users. Along the way the previous significance attached to the presentation of the discipline as a respectable and objective scientific activity has receded, and the importance of being seen to attend to the concerns of users has taken center stage, thanks to the political spotlight focused on the field. The role of ICTs has adjusted as the available technologies have changed and as the pressures on the discipline have to changed. Appropriate technological solutions depend both on what the discipline thinks its problems are and on what the available technologies seem to offer.

New technologies have also provided new opportunities for the discipline to constitute itself *as* a discipline. Mailing lists have provided conduits for disciplinary debate and a public forum for members to keep an eye on where the discipline is at. These lists have helped to normalize the production of electronic resources as a central activity within the discipline. Working through how far the Internet should be adopted as a tool for publishing novel taxonomic work has entailed a reexamination of the relationship of the discipline with its heritage. The advent of new ICTs has offered a cultural disruption that systematics, with its tradition of self-examination, has been well positioned to take up. In discussing whether Web-based consensus classifications should be adopted, how far electronic publishing of new names should be allowed, or how to structure a database to give due credit to contributors, systematists have been discussing fundamental aspects of their communication and credit systems, and the status of their knowledge, their practices, and their users. In this sense new ICTs have presented a therapeutic moment and a significant opportunity for reinvention, even though there is nothing about the technologies themselves to dictate radical change, and, indeed, many of the uses that I have described are notable for the high degree of continuity with previous practices.

ICTs have also provided a branding opportunity and an alternative means of packaging the taxonomic endeavor. They have proved useful as a resource for restructuring temporal qualities of work in systematics, within a pressured funding environment and a consciousness of the need to demonstrate worth. Traditionally systematics has worked on long time scales, with taxonomists focusing on grand life works, and major flora projects taking maybe thirty years to complete. Projects based around databases or Internet resources have been presented as products, even while work is ongoing. A database can be made available on a Web site with only a limited amount of the data that is finally envisaged being present, and an imperfect data set can be presented for user comment and improvement. ICTs have thus been deployed creatively to blur the distinction between ongoing work and finished product, adapting the traditional time scale of taxonomic work to an environment that expects demonstrable results in the short term. The practice of branding projects provides new ways to reach out to audiences for whom a printed flora might not be particularly appealing but who can be persuaded that they might be users of an online resource, with all the connotations of usability and modernity that this evokes.

ICTs have, in fact, been a way for systematics to explore a broad range of potential audiences and to articulate its relationship with those audiences. One key aspect in the current concerns of systematics has been the ability of ICTs to participate in relationships between the herbaria and museums of the more developed nations and those of the often biodiversity-rich but economically disadvantaged nations. In some grander visions ICTs have been proposed as an answer to these inequalities, making expertise and data automatically available independent of geography. In practice, use of these technologies has been a more finely judged and varied activity, as partners in projects begin to work out forms of virtual and material resource that work for them in their practices. Visions of seamless information access for all audiences inform higher levels of debate, but turn into much more nuanced practices on the ground.

Part of the nuancing of practices has revolved around exploring the qualities of material and virtual artifacts and their relationship with one another. Virtual specimens do not, systematists stress, replace their material counterparts. The relationship is more complex, as practicing taxonomists begin to work out just what a virtual specimen could be used for, and as labor is focused on producing virtual specimens that adequately portray valued aspects of material collections. The relationship between virtual and material specimens turns out to be flexible and negotiable, as virtual specimens sometimes do stand in for material ones, and sometimes merely indicate the presence of material artifacts or indeed encourage more use of the material specimens. Specifying the qualities of virtual specimens involves simultaneously talking about what material specimens are good for, conceiving them afresh as repositories of as yet unrecognized information, while at the same time they begin to be seen as excessively fragile and their restriction to particular geographic locations becomes a problem. Remote availability of specimen catalogs and provision of virtual specimen collections augment the existing distributed scientific instrument represented by specimen collections, reinforcing its integrity while never bringing into question the fundamental belief in the specimen collection as a means of reflecting and exploring diversity of organisms "in the wild." Virtual specimens have remained subordinate to material specimens in this regard, and there have been few serious attempts to portray a virtual natural diversity that does not ultimately relate back to the specimen collection.

Virtual collections have been deployed in forms continuous with practices based on use of material collections. Similar observations could be made about taxonomic processes. In descriptions of the bases of phylogenetic analysis, computer packages are treated as manifestations of theoretical propositions, and use of computer packages has come to count as practical experience (Forey et al. 1993; Schuh 2000). Computing has become a natural part of both theoretical and practical work for systematists. In descriptions of taxonomic process computers are portrayed as useful, essential even, but not transformative of the fundamentals. Basic descriptions of how to go about describing new species, such as Winston's (1999) textbook, give little prominence to the role of computers. ICTs have become a flexible component of narratives about taxonomic process, variously foregrounded and backgrounded in accounts. In this regard statements can be viewed as strategic interventions: systematists are aware of the importance of being seen to address the possibilities of ICTs in some quarters, while in other contexts they portray their fundamental activities as unchanging. It is therefore problematic to make authoritative claims about the ability of ICTs to transform systematics. In some representations there appears to have been a radical shift, whereas in other senses the continuity with former manifestations is stressed and computing recedes into the background.

As I described in chapter 1, I have been using cyberscience as a device to foreground ICTs for examination, and though sometimes I have been able to tap into ongoing discussions of these technologies, in other cases I have looked at sites where their use goes relatively unremarked. ICTs play varying parts in narratives told about systematics, depending on the occasion. The investigation by the House of Lords Select Committee on Science and Technology became a site where ICTs were discussed explicitly, and where opportunities arose to depict ICT initiatives as examples of a forward-looking discipline. This not, however, to say that ICTs are necessarily foregrounded in all that the discipline does. Despite suffusing activities and providing a lens for their reimagining ICTs recede into the background in the accounts many institutions and individual systematists give of why they do what they do. For this dispersed community, which is used to exchanging materials and information across geographical distance, electronic communications have become for much of the time like the water they swim in, a largely unremarked way of getting work done. In the missions of the major institutions and in the commitments of individual systematists, ICTs disappear from the starring role they occupy in other kinds of accounts. The focus on developing an understanding of the natural world, spreading knowledge of biological diversity and promoting its preservation take center stage and ICTs become an unremarked means to an end.

In the construction of public pronouncements about systematics, then, ICTs have an ambiguous relationship with change. This ambiguity also works when examining change in relation to working practices. Whereas grander visions of the role of ICTs in systematics have the technology bringing about desirable changes, transforming the discipline into a user-friendly purveyor of timely biodiversity information for the modern age, the reality is, of course, considerably more diverse and achievements are hard won. The gains that are attributed in these visions to digitization are only to be won, it turns out, at the expense of considerable labor. This labor is not always of a kind that is recognized in existing funding schemes, even those that favor digital solutions. Funding agencies are often seen to favor the overtly innovative or the individually packaged solution, and much of the work of digitization is often found simply too long term and too dull to be appealing. In addition, bringing projects into being requires diverse forms of labor that are not always recognized as part of the systematists' skill set. Successful cyberscience is about making politically potent objects, which will be visible to the right people, participate in the right networks and maintain key institutional identities, and also about conducting a practical politics which finds ways for diverse sets of people to identify themselves with and invest themselves in concerted efforts. Getting initiatives to work can involve coordinating an array of interrelated technical, social, and scientific concerns (and indeed working out what might fit into any of these categories), in a form of heterogeneous practice quite familiar to STS (Latour 1996).

Above all, then, the picture of systematics as cyberscience involves systematists as reflexively aware and rounded beings. They are not simply victims of a wider environment that forces them to use ICTs, but are thoroughly aware of cultural currents and pressures, creatively working with them and flexibly shaping their responses to them. Systematics institutions and initiatives engage in a practical politics that involves work across social, technical, and cultural domains to maintain their work as a "doable problem" (Fujimura 1987). While it makes less and less sense, in such a complex, diverse, and active sphere, to ask what the "impact" of ICTs might be, it makes a lot of sense to think of ICTs as a site where contemporary science is being enacted. This is not to say that looking at ICTs produces the complete understanding of science. As I will explore in the final section of this chapter, describing science through its ICTs always has to be a partial account. However, looking at the adoption of ICTs has proved a very illuminating place to think about the complexities of making contemporary science happen.

The overriding picture that emerges from the experience of systematics with ICTs is that of a dynamic process in which ICTs act not as an independent agent of change, but as an evocative resource for thinking about the potential for change and for exploring desires for continuity. Systematics has been able to portray itself as fundamentally unchanged by its experience with these technologies, stressing instead the way it is pursuing its established goals and preserving its heritage. At the same time ICTs have also featured as a major symbolic resource for imagining the discipline as appropriately adapting itself to the expectations and pressures of the modern age. The experience with ICTs is diverse, fragmented, and highly specific to context, and encompasses both the sense that the discipline is adapting and that it remains fundamentally the same. ICTs have offered systematics new ways to imagine its identities, its ambitions, and its histories. At the same time, however, the experience of novelty that these technologies offer has been suffused with a sense of continuity. As much as ICTs give systematists new ways of imagining themselves, practitioners have been drawn to narratives that connect with other more familiar stories they tell. New technologies provide therapeutic moments with which to refashion goals and identities, but systematists seem to be drawn to forge connections between past and future rather than opting for the more radical new beginnings envisaged by some commentators.

Cyberscience, Change, and Specificity

When people use computers it is because they find this a socially meaningful thing to do. We expect to be able to talk about what we are doing in ways that make sense to people around us, and in the process we weave our use of technologies into other aspects of our lives. In this regard scientists are no different. Sometimes the kind of activity they are engaged in may seem esoteric to outsiders, and their descriptions may use special vocabularies and concepts. The basic premise, however, remains the same: sustainable activities will be those that scientists can meaningfully represent to themselves and to others within the spectrum of ongoing activities and developing priorities. There seems, then, to be a lot of continuity between science and the rest of society in the ways that new technologies are interpreted, appropriated, and adapted. The descriptions of systematics as a cyberscience in this book have shown systematists taking ICTs and turning them into a socially meaningful component of their endeavors. There is a specificity to the adoption of ICTs: they are deployed in ways that are quite contextually particular, and for each group they are interpreted afresh. There is, however, a common thread that cuts across domains, concerning this particular set of technologies that have acquired a cultural potency as evocative resources for thinking about change and continuity. In this section I will therefore explore connections with other frameworks for understanding ICTs and imagining in domains outside science, as a means to thinking about how far the dynamics identified here for cyberscience are specific to science.

The contemporary preoccupation with technologies as models for imagining change has been highlighted by Barry, who talks about the existence of a technological society within which "specific technologies dominate our sense of the kinds of problems that government and politics must assess, and the solutions that we must adopt. A technological society is one which takes a technical change to be the model for political invention. The concept of a technological society does not refer to a stage in history, but rather to a specific set of attitudes towards the political present which have acquired a particular contemporary intensity, salience and form" (Barry 2001: 2). This form of society is described as qualitative instead of chronological, indicative of a particular set of preoccupations rather than a stage of development. Cyberscience can be thought of as a similar order of concept and is also substantively connected in that it draws on similar sets of evocations. As we have seen in systematics, and in talk of e-science, cyberinfrastructures, and collaboratories, ICTs have become a significant preoccupation within science, offering up a potent set of ways to portray past and present and to project future visions. There is a continuity to the expectations that surround ICTs in the political domain and in science. Just as Kling and Iacono noted in the computerization movements they observed (1988, 1996; Iacono and Kling 2001) there is a publicly constituted self-evidence to the deployment of ICTs that cuts across domains. These technologies provide visions for directions in which to push diverse aspects of contemporary life, and in this sense cyberscience may be seen as a localized manifestation of a dynamic that is not confined to science alone.

To pursue this point about the continuity of the dynamics of cyberscience with those experienced in other cultural domains I will take a comparative framework from another ethnography, conducted in quite different territory. I have used the notions of co-construction and imagining to stress that deployment of ICTs, for systematics, entails a dynamic recasting of both identity and technology. Miller and Slater (2000) in a similar move see the Internet as a resource that is deployed in people's ongoing imagining of themselves. Miller and Slater focus specifically on the Internet as experienced by Trinidadian culture. Although in the current study I have taken the Internet as part of a wider complex of technologies that formed the focus of my fieldwork, the Internet forms a key evocative component of this complex of technologies. It thus seems appropriate to take Miller and Slater's observations on the Internet as a template for thinking about how far the processes involved in realizing systematics through ICTs are specific to science. Miller and Slater (2000) offer a set of concepts as a basis for comparing Internet adoption across domains. Specifically they suggest that Internet adoption entails four key dynamics: objectification, mediation, normative freedom, and positioning. These dynamics provide a useful way to think through the specifics of the form of cyberscience that stems from the engagement of systematics with ICTs, and give a further source for reflection on the relationship between cyberscience and change.

In terms of objectification, Miller and Slater (2000) argue that people are using the Internet as a means of being who they are and exploring who they could be. The dynamics of objectification are particularly striking in the experience of systematics with ICTs. I have used the notion of co-construction to highlight the extent to which ICTs, as much as they offer new possibilities for systematics and have provided a site for imagining what the discipline should be, have also themselves been imagined in particular ways as appropriate to what the discipline wants to be. ICTs have provided a site for forging versions of the discipline that feel, for its practitioners, appropriate both to contemporary pressures and to commitments to continuity with key aspects of disciplinary heritage. For individual systematists too, ICTs have in diverse ways offered possibilities for being themselves in new forms, whether in becoming distinguished through a particular set of ICT-related competences or activities, or in portraying themselves on the stage offered by mailing lists and Web sites.

The dynamics of mediation relates to the way the Internet is constituted as a medium in relation to other media and understandings developed of how it can and should be used. This notion provides a way to think through the processes through which ICTs are disaggregated by systematists and made meaningful as very specific forms of media in relation to particular activities. The Taxacom mailing list thus becomes a medium for a form of disciplinary engagement through the practices of individuals using it. Various forms of online resource become components of working practice, but this is intricately linked with developments in ideas about what it is to work effectively with objects and also with developments in their manifestation within communication regimes and material culture. Issues as mundane as how fast an institution's Internet connection might be, or where the computers are placed in relation to the specimen collection, help to shape the potential of ICTs as specific media for specific working practices. At the same time, however, these dynamics play out in relation to broader expectations about what particular groupings of technologies offer: as Miller and Slater (2000) argue, "it is important to understand the Internet as a symbolic totality as well as a practical multiplicity."

The notion of a dynamics of normative freedom points to the concern that although the Internet has often been understood through a lens of liberatory potential, many instances of use close down on the possibilities and enact connections with previous ordering frameworks. This dynamic can be seen particularly clearly for systematics, where glimpses of the possibility of a seamless sphere of biodiversity information suffuse the visions for what the Internet could be, and yet various concerns close down on the radical potential and preserve prior notions of property, credit, authority, and institutional territories. Ideas about the potential for ICTs to provide accessible information on biodiversity directly to users become filtered through concerns about the need to preserve order and to attend to the various structures that continue to prevail across the discipline. ICTs do not sweep away the past but engage with it in dynamic fashion.

The final dynamic that Miller and Slater point to is the concern with positioning, and the reflexive engagement with relations to other that is enhanced by and enacted through the use of the Internet. For systematics, ICTs have been a way of developing a positioning for a stage that has increasingly come to seem global. There has long been a sense of systematics as a transnational practice involving diverse forms of connection between sites. ICTs have been used in ways that enhance this sense of systematics, and particularly specimen collections, as part of a transnational infrastructure for the study of biological diversity. At the same time, ICTs have offered a way of positioning institutions in relation to their competitors and of portraying the discipline of systematics to other sciences and to its funders as an appropriately forward-looking contemporary science.

Miller and Slater provided a lens for exploring ways in which Trinidadians made the Internet their own in the context of the very particular political and economic circumstances they inhabited. It became a way of both staying the same and moving into the modern world. I find the sets of dynamics that Miller and Slater (2000) identify a useful way to think about the processes by which scientific disciplines may be able to claim ICTs as their own and at the same time find themselves anew within them. Although the outcomes are highly specific, then, the dynamics through which they are achieved may be recognizable in abstracted form across domains.

In chapter 2 I introduced the notion of the technology in practice (Orlikowski 2000), denoting that technological capacity could be seen as a thoroughly situated notion developed uniquely in particular contexts of deployment. The idea of co-construction, as applied to the tools of science, offers us a way of understanding that the technology in practice is achieved in conjunction with science in practice and with particular disciplines in practice, just as we might say that for Miller and Slater's (2000) study the Internet became "Trinidad in practice" and that looking at Trinidad enabled them to study an "Internet in practice." Cyberscience is a form of science realized through and imagined in relation to a particular set of technologies. Rather than being a phenomenon specific to science, cyberscience seems to be representative of a form of imagining that resonates with Barry's (2001) technological society and is plausibly understood in terms of the kind of dynamics that Miller and Slater (2000) identified in the Trinidadian Internet. At the same time as cyberscience disaggregates into diverse phenomena for different disciplines and research groupings, it also connects with similar dynamics across diverse domains beyond science.

Cyberscience as an Object of Policy

If cyberscience is about imagining change and continuity rather than implementing transformation, there are serious implications for those who invest money in it and encourage its development. At the outset I positioned this book as a contribution to the understanding of contemporary scientific culture that could have something to say in a policy context. The prospect of gaining more return for the funds expended on science by providing and promoting better information and communication infrastructures has animated much policy debate and been the focus of considerable investment in recent years. I set out to explore one case study with the hope of engaging in that debate and considering how far some of the expectations that motivate it might be realized. In this section I examine some of the policy recommendations that might be made in the light of this case study, focusing particularly on the observations I made in previous sections about the processes of co-construction and imagining that surround cyberscience and their implications for the inherent specificity of the outcomes.

It will have become clear that the equation of information and communication infrastructures with efficiency is somewhat misguided. Efficiency means very different things for different disciplines, and the forms in which it might be assessed and the relevant time scales for doing so will vary widely. For systematics, the desire to be seen as efficient animated some institutional willingness to embrace new technologies, and motivated (or made funding available for) some projects aimed at consolidating information across institutions. Nonetheless, for many projects efficiency was at best a far-off goal, never likely to be measurable in any plausible way. The archival nature of systematics collections, and the corresponding focus on virtual specimen databases as reflective of the valued qualities of the archive, means that the time scale of the payoff is likely to be very long, if it ever becomes discernible. A range of motivations, including the fragility of material collections, the political pressure to share, and the existing orientation toward loans and visits, makes digitization of collections a meaningful thing to do, but efficiency is a poor shorthand to use for this array of interlinked rationales.

The prospect of efficiency gains is also insufficient to produce change in the face of other cultural pressures and sources of inertia. Godfray's (2002a,b) proposals for Web-based taxonomies aimed at efficiency by doing away with the need to consult the legacy literature and providing a one-stop source of information for users. His initial proposals lacked cultural viability in the face of the strong orientation toward the literature and the political sensitivities of systematists regarding centralized and potentially elitist solutions. A more culturally viable manifestation of efficiency to deal with the legacy literature problem may prove to be digitization of the legacy literature, but this again depends on some very specific qualities of that literature unique to taxonomy. Where the policy goal is increased efficiency in the science system, ICTs are an imperfect instrument, and it becomes important to ask what efficiency means for the particular community concerned. Rather than a straightforward formula of "more science for less money," efficiency has to be assessed in the context of the ways that a particular community works and the kinds of change that will be culturally viable for that community to adopt.

E-science and cyberinfrastructures are often associated with collaborative work and with the ability to address questions across disciplinary boundaries and on a grander scale than previously possible. Again, the experience of systematics makes clear that the technologies do not possess an independent agency to make people collaborate or address new questions. They can prompt new directions, but only insofar as these make sense in the wider political and funding climate and as applied to the participants' understanding of priorities. It is clear from the experience of systematics that grand visions are interpreted through on-the-ground experience, and they often become diluted, appropriated, and diversified so that it is difficult to recognize the original vision in the final products. Nonetheless, there is a valuable payoff in terms of the occasions for reflection, debate, and recognition of change that these grand visions offer. Greater recognition could be given in policy circles to this kind of opportunity: funding for new programs and initiatives could usefully be accompanied by encouragement to organize occasions for reflection, both within and across communities, on prospects for change and experiences of appropriation.

One significant aspect of reflexivity that ICTs have occasioned for systematics has been the chance to think about audiences and users. Designing a new technology routinely involves thinking about the users for that technology, and this disruption to the usual genres of taxonomic reporting has offered opportunities to think about who might want taxonomic information, in what forms, and what their expectations might be. This is a valuable reflexive payoff for the systematics community, beyond any specific practical outcomes of particular projects. However, there is an accompanying frustration in that knowledge of users is sparse and imperfect. Much second-guessing of users' needs and practices goes on in the absence of direct contact. Funding instruments could address this frustration directly by making available money to involve users, motivating users to identify themselves as such and to get involved in shaping the information resources they need. Darwin Initiative funding provides a model for this kind of engagement, focusing as it does on forging partnerships to overcome the taxonomic impediment by sharing expertise with those who have an identified need for it. This funding model could be extended more broadly to initiatives in new infrastructures for science.

Some further observations about the viability of initiatives as observed in systematics may have a broader applicability to e-science initiatives. It is clear from the experience of systematics that viable projects will have to consider far more than simply a requirements analysis focused on the qualities of data and the practices of working with them. In the first instance, requirements in science, as in other domains, do not necessarily preexist the project design phase, but emerge through the course of design. Where data are not "born digital," participants may be thinking for the first time about which qualities they value in material artifacts in order to represent them in virtual form. This will be a complex process, involving many uncertainties and addressing some fundamental issues of disciplinary practice through a dynamics of co-construction. In such circumstances digitization will involve considerable labor, in varied forms that both invoke established expertise and develop new skills. Funding and credit structures will need to make this kind of labor meaningful and reward it appropriately. Leaders of viable projects will need to be expert practical politicians, attending to credit and reputational structures, funding frameworks, institutional landscapes, and understandings of intellectual property that, again, may not have been explicitly specified prior to the existence of the project.

I have assembled some practical points where a policy for sustainable e-science might usefully direct attention, learning from the experience of systematics. However, I have positioned the case study of systematics as a source of questions to ask in relation to e-science, rather than a cause for making absolute recommendations, because a key part of my argument is that experiences with ICTs will be highly specific to particular communities. On this question of specificity, it is salutary to reflect on how different the experiences of systematics have been from those in genomics. The use of databases as a research resource and as an appropriate place to lodge one's data has become routine practice in genomics (Brown 2003). Understandings of the potential of databases as a means of storing and making raw research data accessible have formed the focus for new communication regimes to develop (Hilgartner and Brandt Rauf 1994; Hilgartner 1995). ICTs have also acted as a resource for imagining the future of genomics, in this case forming a grand vision for a radical transformation to a new scale of biology (Lenoir 1998b). It could be argued that genomics is also a cyberscience, in that its contemporary experience is suffused with ideas about ICT-enabled routes to desirable outcomes. There are, however, key areas where the experience of genomics has been very different.

For genomics, the relationship of cyberscience with the heritage of the discipline contrasts greatly with the situation I have described for systematics. Whereas systematists negotiated a future vision in dialogue with their heritage, and were suspicious of calls to sweep it away and move forward, genomics seems rather to have appropriated the grand vision and used this acceptance to negotiate the institutional, technological, and social requirements to realize the vision. A part of this, again, depends on the particular imagining of the potential of genomics as a medical tool and economic asset that came to dominate political thinking in the 1980s, such that governments were concerned not to be left behind and consequently were prepared to invest large amounts of money in making grand visions into viable reality. Another reason for the readiness to embrace a vision of ICT-enabled transformation in genomics could be the consciousness of rapid change in the discipline revolving around the advent of DNA-sequencing, making it easier to imagine that now is the time for a radical departure from past practices.

The institutional landscape of genomics is quite different from that of systematics, as is the relationship between data and material culture according to the established working practices of the discipline. Whereas in systematics data are always tied back to material collections through type specimens, in genetics the data (a gene sequence, for example) are more readily seen as independent objects, and thus data repositories have less of a relationship with material artifacts and the institutions where they are kept. Some quite distinctive institutional forms arose in genomics focused on provision of data and data analysis services (Balmer 1995, 1996; Glasner, Rothman, and Travis 1995; Glasner, Rothman, and Yee 1998). This different relationship with material artifacts and institutional arrangements has made it easier for large data repositories to be instituted within genomics and to become resources to be inserted into working practices in laboratories across the world. A centralized data repository in genomics can aspire to become the data source of choice, whereas in systematics the scope of a material collection and the institutional desire to protect resources limits the extent of individual data repositories and focuses attention on efforts to provide portals that offer seamless access to diverse resources.

Within genomics, as in systematics, electronic media have become an accepted part of the informal communication practices of the discipline. A variety of forms of discussion forum and announcement list populates and enacts the discipline, along with widespread use of email for coordinating collaborative work. Just as in systematics many participants saw the Taxacom list as continuous with their experiences of the discipline in other media, so in genomics electronic forums have developed in a form that feels for participants like a continuation of laboratory space (Hine 2002). Informal communications via electronic media have therefore for both disciplines provided a way of coming together and realizing aspects of practice and disciplinary identity and have become seamlessly incorporated into the existing array of communications, although again the outcomes are thoroughly specific to each disciplinary culture.

In regard to formal publishing, the relationship between databases and literature has again been worked out in forms very specific to genomics. In particular, the key negotiation for genomics was the relationship between journals and databases, revolving around the articulation of the need to preserve the quality of data made accessible through databases (Hilgartner 1995). The formation that emerged cemented the relationship between journals and databases by requiring authors of journal articles as a condition of publication to lodge the corresponding data in a database. This depends on a very specific negotiation of positions, and similar solutions have proved less viable for brain-imaging repositories (Beaulieu 2001, 2003). Within systematics, the key negotiation seen as crucial in safeguarding quality revolved not around journal publishers as the gatekeepers of peer review but focused instead on institutions, as the holders of specimen collections, and individual systematists, as the acknowledged experts for a particular group of organisms. The concerns not to let poor quality data into the public domain were, for systematics, a continuation of existing practices of curation. Systematics also adds an additional set of considerations through the very particular status of its legacy literature and the system of nomenclatural codes that provide key sites of negotiation for considering the viability of new publishing forms using ICTs. Both systematics and genomics have found themselves thinking about what data should be made available to whom. Whereas systematics had to orient to geographically dispersed users who must, for political reasons, be addressed, genomics had a distinctive focus on commercial users and producers of data with whom to negotiate workable practice (Groenewegen and Wouters 2004).

Shaped thus by their distinctive political status, institutional structure, relationships of data to material artifacts, and communicative practices, systematics and genomics have worked out very different responses to the visions that ICTs evoke. ICTs have an interpretive flexibility such that we can expect differences between disciplines in the way that ICTs are interpreted and the visions through which their possibilities are enacted (Bohlin 2004). There will also, though, be considerable differences within disciplines in the way that emerging technologies are institutionalized and inserted into working practices. Within genomics, for example, not all uses of databases are readily interpreted as instances of communication regimes;

they can instead be instituted as research tools on a much smaller scale, although they can still involve spatially and socially complex patterns of collaboration and connection (Hine 2006b). Given the levels of inter- and intradisciplinary diversity in the enactment of ICTs, it therefore becomes clear that, potent as ICTs are as a resource for imagining change in science and rearticulating past practices, the dynamics are complex and contingent and the outcomes diverse. This is not to say that each disciplinary deployment of ICTs should be seen as an independent response to some free-standing vision which each consumes independently. Disciplinary responses are mutually visible and mutually elaborated: as chapter 3 described, genomics figures in some descriptions of what systematics should aspire to be. Diverse disciplinary responses to ICTs therefore build on and respond to one another, emerging as a complex of interreferring yet distinctive practices that connect with broader cultural dynamics.

Viewed in this way, cybersciences, e-sciences, or cyberinfrastructures will inevitably be diverse in nature and specific to the disciplinary cultures within which they are enacted. A wholesale change toward a more efficient, collaborative, or larger-scale science is unlikely as each discipline, subdiscipline, or research grouping works out what makes sense for its concerns. Such change as is apparent will often be the result of improvisations and local adaptations, will orient to specific political or policy pressures, and may occur over far longer time scales than most policy instruments allow for. By trying to get on with their work as they understand it, and by understanding their work through the technologies that are available to work and think with, scientists may enact change, but this will be in directions that they are able to make meaningful to themselves and to those around them. This focus on specificity means that general-purpose solutions may prove unpopular, and the reusability of investments in infrastructure development for individual communities may be limited. Economies of scale from the reuse of infrastructures across communities may therefore be difficult to realize: but the process of trying to work out what might and might not fit could still be very fruitful for the communities concerned.

Cyberscience and STS Fieldwork

Using a connective ethnographic approach that implements cyberscience as a notion to pull together observations about experience with diverse technologies has afforded me the opportunity to develop a multifaceted understanding of a scientific discipline, one that, through this culturally potent category of technology, links observations across the often artificially separated domains of working practice, institutional landscape, communication system, and political context. These domains prove to be richly interwoven in the experience of working with ICTs and in the constitution of disciplinary identity. By following ICTs in practices and narratives across diverse sites and between online and offline I was able to explore ways that meaning-making cuts across domains and connects them. Science and many other cultural phenomena are conducted across diverse geographic and virtual landscapes, and it seems likely that other ethnographers will want to use this kind of approach to attend to the varied textures of these interconnected landscapes. I therefore use this closing section to reflect on the mixed blessings of this approach to ethnography of cyberscience. In particular I have missed out on some of the purchase that detailed engagement with day-to-day working practice would have afforded, and the resulting potential losses are worth evaluating in relation to other approaches to ethnography of science.

I have come to my version of the ethnography of cyberscience by taking an approach that values finding out how the world is structured from the perspective of participants and pursuing their meaning-making practices. In particular, I set out to explore the conjunction of a particular disciplinary culture with a particular set of technologies, and to examine the issues that the technology brings to the fore for that discipline. I set out to produce an ethnographically informed study of science in the tradition of science and technology studies. In the execution, however, as I now survey what I have done, I have produced a study that sits somewhat oddly with the pioneering and influential ethnographies of science. It is implicit in my account that science is amenable to sociological investigation, and that social concerns sit at the heart of what systematists do. I have not, though, produced an account that focuses on particular products and processes of systematists' work; nor have I outlined how various sociological dynamics informed the knowledge that came to be produced and valued by the community. In order to engage with these epistemic issues it might have been fruitful to have moved around less, and to have focused in more depth on situated practices of production and consumption. The field site and fieldwork practices for this set of issues would have been somewhat different, and for this the strategies of laboratory ethnographers remain highly salient. In particular, an appropriate focus for that fieldwork might have been to follow a particular small research grouping wherever their work took them and to focus on a specific scientific problem that preoccupied them, instead of moving around in search of a discipline as I did.

Looking at cyberscience through a connective approach I found diverse notions of influence and autonomy carving up the field of experience, leaving it unclear whether discipline, institution, individual, or initiative were the appropriate unit of analysis, and rendering any one of them inadequate on its own. I focused on a discipline as my basic unit of analysis, but in doing so I inevitably ignored other connections that could have been drawn and sacrificed depth of engagement with knowledge construction processes. All fieldwork entails making situated decisions and compromises: it is in the nature of any approach where the researcher is the research instrument, adapting to the circumstances found in the field. A consequence of taking an overtly connective approach, however, is that the nature of the field site becomes a particular source of anxiety, innovation, and a topic all of itself. Instead of respecting boundaries the focus turns to connections, developing a sensitivity to the meaning that various forms of connection have for those involved, and this entails a consciousness of incompleteness and of potential connections unpursued.

Within a "connective" approach to ethnography a connection can take many forms, such as a Web link, a mention in a report, a set of participants in a project, or a recommendation from an interviewee. As experience of the field develops so too does some confidence about judging which connections are important and which are more trivial or accidental, and security about having a domain of interest about which something sensible can be said (if not a conventional field site) increases. In the telling of my account I have therefore presented it with a certain amount of confidence as to how the discipline *is*, based on my experiential understanding; but at the time of conducting the fieldwork everything seemed shifting and tentative, and it took time for the various connections to begin to align with one another and become plausible as ways to account for the discipline's preoccupations and practices. Doing this kind of ethnography is about developing an understanding of what is going on so as to find the next appropriate place to go to.

Understanding is about immersion, so there is a need to develop an appropriate researcher presence for different media and to engage with the people encountered via those media. I found that immersion was helped by identifying myself carefully, especially in virtual spaces, and by developing a Web site that showed an appropriate persona for online contacts and potential interviewees to view. Engagement through the media of interest became a learning experience in its own right: it became interesting not simply to use email to contact potential interviewees or to research them through their Web sites, but also to reflect on the media employed and the extent to which these were valued components of the disciplinary landscape. Whereas an ethnography based on a single laboratory tends to rely on building sustained relationships, the connective approach asks for a continued focus on making new engagements, which can, in their own right, be used to prompt reflection on the variety of forms of relationships that make up contemporary science (Heath 1998; Heath et al. 1999).

Compared with my previous attempt at virtual ethnography, in this project I pursued similar principles but found that they took me to different places. I stuck to my belief that ethnography could both engage with computer-mediated communication as a site of meaningful social interaction and also pursue the diverse connections that helped those interactions to make sense: an approach captured in the slogan that the Internet is both culture and cultural artifact. In seeking to understand a scientific discipline's adoption of ICTs, I found that I was drawn to follow connections into aspects of material culture, policy, and practice that were not manifested online and were certainly not amenable to a wholly online ethnographic approach. Whereas the last time I celebrated the dual existence of the Internet as culture and cultural artifact largely through online engagement, this time I explored in much more depth how that artifactual status was achieved and sustained across diverse sites. I offer an account that reflects the varied orientations that being a contemporary taxonomist entails, identifying diverse factors that shape the role of ICTs as a manifestation of and influence on the disciplinary landscape. This kind of attention to the lived reality of scientific experience as it relates to ICTs is, I would suggest, a valuable partner to the detailed ethnography of knowledge-construction processes. The connective approach offers a complementary way to comprehend science as a lived experience that orients to and reflects diverse sociotechnical developments.

This project was not a conventional ethnography but a more mobile project that paid correspondingly less attention to the detail of scientific practice and more to the varied sites in which the discipline was manifested. A "laboratory ethnography" style of investigation would undoubtedly have gained in depth of understanding of how taxonomic work is done, but, I would argue, would have missed out on some factors very important in understanding contemporary scientific practice for the particular purpose that I began with and for the particular technologies that were my focus. To understand ICTs as cultural artifacts I found it useful to think of cultural meaning as a highly complex and multilocated phenomenon. Users and developers as reflexive beings are conscious of working with and shaping the cultural status of the artifacts they employ. The grand and the mundane, the practical and the political are not separate levels of analysis, but are interwoven, not as different scales, but as differing orientations. ICTs act not as an agent of change in themselves, but as an occasion for reflection, realignment, specification, ordering, adjustment, strategic thinking, and vision. They act as cultural disruptors, providing occasions for shifts to be experienced and articulated. This potency depends on the ideas of progress and transformation which they have come to represent. Mobile methodological solutions help to address this breadth of scope and heterogeneity of connection even though the shift to mobility involves losses as well as gains.

It was important in the study I conducted to travel, to traverse geographic and virtual space and explore connections between them, to attend to institutional landscapes and the working environments of individuals, and to explore policy documents alongside contemporary practice. My study acquired an understanding of the varied textures of lived experience in contemporary systematics seen through the lens of its engagement with ICTs, and, in ethnographic tradition, the way that I did this was by moving through that territory myself, gaining an experiential understanding through my immersion within it and my interaction with those I encountered in it. The resulting picture of science is thoroughly partial and, of course, thoroughly personal. As I explored in relation to virtual ethnography (Hine 2000), following connections is an ethnographic strategy that provides insights into meaning-making structures at the expense of aspirations to the completeness or security that comes from adherence to bounded notions of field sites. Connective ethnography is a potent way to understand contemporary science, but it is not an automatic route to a truer picture of what science is. As with all forms of ethnography, it stands as an intervention that is shaped by the preoccupations of the ethnographer in dialogue with an emergent field, and it ultimately acquires its adequacy in its engagement with audiences and the questions they ask of it. One interviewee asked me, at the beginning of the interview, whether I was "one of those mad cultural relativists." I hope he would now agree that, even if I deserve the description, we mad cultural relativists can sometimes make sense.

Appendix: Messages Sent to the Taxacom List Asking for Input

Message 1

I am mailing the Taxacom list to ask whether list members would be willing to help with some research that I'm currently undertaking. I am holder of a research fellowship from the UK's Economic and Social Research Council to look at information in contemporary science. I'm currently researching the role that information and communications technologies play in systematics—looking at high level policy documents and initiatives as well as the more direct experiences of those curating collections and working in taxonomy. I'm using systematics as a case study within an overall aim to give a grounded and sociologically sophisticated perspective on information and communications technologies in science more generally to guide policy initiatives.

One of the aspects that is really important to look at is use of computers to communicate amongst scientists. And this is where my interest in the Taxacom list comes in. I've been using list archives to explore some of the issues that have been important to systematists over the years. Where I've wanted to quote from a message I've contacted the author directly, and found that people have been really helpful. But now I have some broader questions about what the list means for systematics that I'd be grateful for your help with. My questions (or some of them), are these:

How important is this list for a practicing systematist today? Would you miss it? What would taxonomy be like without it?
How far do the kinds of issues discussed on the Taxacom list reflect the concerns of the discipline more broadly? Is there an excessive focus on particular kinds of issues? Do others get missed out?

Have you posted messages to the list, either to start a topic or respond to one? What was your experience like—did you find it helpful, enjoyable, or neither?
How many of the people who contribute to the list do you know from other contexts? Have you met many of them face-to-face?
What other lists do you belong to? How does this list differ?
I'd be particularly interested to hear from anyone who never or rarely sends messages to the list, but still finds it useful—what benefit do you get from the list? Do you know colleagues in taxonomy who don't subscribe to the list, and do they miss out?

I'm a qualitative researcher, so this isn't a survey. The questions are starting points to which I'd love to hear responses—or tell me if you think I'm asking the wrong questions. It wouldn't be fair to clutter up this list, so please do respond to me personally. I will, however, aim to report back to the list in a few months with the outcomes. I'll keep names and identifying information confidential unless otherwise explicitly agreed. In the meantime there is more about my background, and some relevant papers I've written at http://www.soc.surrey.ac.uk/christine_hine.htm. Apologies for sending such a long message!

Many thanks,
Christine

Christine Hine
Department of Sociology
University of Surrey
Guildford, Surrey, GU2 7XH, UK
c.hine@surrey.ac.uk or christine.hine@btinternet.com
http://www.soc.surrey.ac.uk/christine_hine.htm

Message 2

I posted some questions recently to ask for input into research on the taxacom list, and what its significance is for work in systematics. Many thanks to everyone who has taken the time to respond to my recent questions. For anyone who missed it, here is the link to the original message: http://list serv.nhm.ku.edu/cgibin/wa.exe?A2=ind0405&L=taxacom&D=1&O=D&P=5064. I'd still be very interested in further responses—I have certainly by no means exhausted the range of possible variability, since each response I've received has given something quite different to think about.

I will, in due course, post a link to a report on this research. In the meantime, I have some further questions that have arisen from looking at the list and from responses to the initial questions.

There are clearly a lot of women working in systematics, and indeed subscribed to the list. But few seem to post messages. Why might this be? Is it a problem?
Some people clearly spend a lot of time contributing to the list. What motivates you?
Apart from the public replies sent to the list, do messages generate private email correspondence? Do many of you respond privately to queries, rather than to the list?

I'd be grateful for any feedback on these or on the original questions—as before, direct to me and not to the list, to avoid clogging up the list.

Best wishes,
Christine

Christine Hine
Department of Sociology
University of Surrey
Guildford, Surrey, GU2 7XH, UK
c.hine@surrey.ac.uk
http://www.soc.surrey.ac.uk/christine_hine.htm

Notes

Chapter 1

1. Systematics is generally used as the overarching term, to denote the study of the relationships between organisms. Taxonomy lies within this field as the practice of classifying and naming organisms. In this book the two are used interchangeably: but see chapter 3, note 1 for some reservations about this decision.

Chapter 2

1. Woolgar et al. (2005) talk of an "entangled and on-going assessment and questioning of STS engagement, utility and accountability" which is manifested in an array of recent workshops including those organized by themselves at the Saïd Business School in Oxford in 2004 and 2005 on the theme "Does STS Mean Business," and to which can be added, more recently, workshops on "Unpacking 'Intervention': Action-Oriented Science and Technology Studies" organized at in May 2006 in Århus by Casper Bruun Jensen and Teun Zuiderent-Jerak and also a workshop on "Middle Range Theory and STS" organized by Brian Balmer and Sally Wyatt in Amsterdam in April 2005. Each occasion, in a different way, addressed the notion of an STS engaged with domains of use and examined resulting tensions around the negotiation of adequacy for STS accounts.

Chapter 3

1. We can see that the decision made in this book, to use systematics and taxonomy interchangeably, is probably quite naive in the light of Hedgecoe's (2003) analysis. Although this is not the topic of the current discussion, it is likely that the connotations of systematics and taxonomy do differ in important ways for both participants and onlookers.

Chapter 5

1. Collembola are a group of wingless arthropods, often soil-dwelling, with approximately 7,500 known species (Bellinger, Christiansen, and Janssens 1996).

2. Bacteriology and virology have been more open to these changes: altered techniques and major recent advances have particularly diminished the usefulness of the older literature in these fields, and the problems of synonymy in bacteria were particularly complex.

3. A wiki is a form of Web site in which any person (or often, any registered person) is able to edit and add pages. The wiki is therefore open to collective construction, making it a useful tool for contributing comments to a development project.

References

Aborn, M. (1988). Preface. *The Annals of the American Academy of Political and Social Science* 495 (Telescience: Scientific Communication in the Information Age): 10–13.

Agar, J. (1998). Screening science: Spatial organization and valuation at Jodrell Bank. In *Making Space for Science: Territorial Themes in the Shaping of Knowledge,* ed. C. Smith and J. Agar, 265–280. Basingstoke: Macmillan.

Agar, J. (2006). What difference did computers make? *Social Studies of Science* 36(6): 869–907.

Aguilar, J. L. (1981). Insider research: An ethnography of a debate. In *Anthropologists at Home in North America: Methods and Issues in the Study of One's Own Society,* ed. D. A. Messerschmidt, 15–26. Cambridge: Cambridge University Press.

Allison-Bunnell, S. W. (1998). Making nature "real" again: Natural history exhibits and public rhetorics of science at the Smithsonian Institution in the early 1960s. In *The Politics of Display: Museums, Science, Culture,* ed. S. MacDonald, 77–97. London: Routledge.

Allkin, R., and F. A. Bisby (eds.) (1984). *Databases in Systematics.* Systematics Association Special Volume 26. London: Academic Press.

All Species Foundation (n.d.). FAQs. Http://www.all-species.org/faq_response.html/. Accessed 15 December 2004.

American Association of Museums (2003). AAM position statement: University natural history museums and collections. Http://www.aam-us.org/pdf/univcollstatement.pdf/. Accessed 9 January 2004.

Amit, V. (ed.) (2000). *Constructing the Field: Ethnographic Fieldwork in the Contemporary World.* London: Routledge.

Attewell, P., and J. Rule (1984). Computing and organizations: What we know and what we don't know. *Communications of the ACM* 27(12): 1184–1192.

Backes, P. G., K. S. Tso, and G. K. Tharp (1999). The Web interface for telescience. *Presence-Teleoperators and Virtual Environments* 8(5): 531–539.

Bailey, C. W., Jr. (2005). *Open Access Bibliography: Liberating Scholarly Literature with E-Prints and Open Access Journals.* Washington, DC: Association of Research Libraries. Http://www.escholarlypub.com/oab/oab.pdf/. Accessed 15 December 2006.

Baird, D. (1993). Analytical-chemistry and the big scientific instrumentation revolution. *Annals of Science* 50(3): 267–290.

Baird, D., and T. Faust (1990). Scientific instruments, scientific progress, and the Cyclotron. *British Journal for the Philosophy of Science* 41(2): 147–175.

Balmer, B. (1995). Transitional science and the Human Genome Mapping Project Resource Centre. *Genetic Engineer and Biotechnologist* 15(2,3): 89–98.

Balmer, B. (1996). Managing mapping in the human genome project. *Social Studies of Science* 26(3): 531–573.

Barjak, F. (2004) On the integration of the Internet into informal scientific communication. Series A: Discussion Paper 2004-W02. Solothurn University of Applied Sciences Northwestern Switzerland. Http://www.fhso.ch/pdf/publikationen/dp04-02.pdf/. Accessed 15 December 2006.

Barjak, F. (2006). From the "analogue divide" to the "hybrid divide": The Internet does not ensure equality of access to information in science. In *New Infrastructures for Knowledge Production: Understanding E-Science,* ed. C. Hine, 233–245. Hershey, Penn.: Information Science Publishing.

Barry, A. (2001). *Political Machines: Governing a Technological Society.* London and New York: Athlone Press.

Beach, J. (1992) Taxacom mailing list news. Biological systematics discussion list (taxacom@harvarda.bitnet). Archived at http://listserv.nhm.ku.edu/archives/taxacom.html/. Message date 2 November 1992.

Beaulieu, A. (2001). Voxels in the brain: Neuroscience, informatics, and changing notions of objectivity. *Social Studies of Science* 31(5): 635–680.

Beaulieu, A. (2003). Research woes and new data flows: A case study of data sharing at the fMRI Data Centre, Dartmouth College, USA. In *Promise and Practice in Data Sharing,* ed. P. Wouters and P. Schröder, 65–88. Amsterdam: NIWI-KNAW. Http://www.virtualknowledgestudio.nl/staff/anne-beaulieu/documents/the-public-domain-of.pdf/. Accessed 15 December 2006.

Beaulieu, A. (2005). Sociable hyperlinks: An ethnographic approach to connectivity. In *Virtual Methods: Issues in Social Research on the Internet,* ed. C. Hine, 183–198. Oxford: Berg.

Beccaloni, G. W., M. J. Scoble, G. S. Robinson, A. C. Downton, and S. M. Lucas (2003). Computerising unit-level data in natural history card archives. In *ENHSIN: The European Natural History Specimen Information Network,* ed. M. J. Scoble, 165–176. London: Natural History Museum.

Bellinger, P. F., K. A. Christiansen, and F. Janssens (1996). Checklist of the Collembola of the world. Http://www.collembola.org/. Accessed 31 December 2003.

Benschop, R. (1998). What is a tachistoscope? Historical explorations of an instrument. *Science in Context* 11(1): 23–50.

Berendsohn, W. G. (2000). Resource identification for a biological collection information service in Europe. Results of the Concerted Action Project funded by the European Commission, DG XII, within the EU Fourth Framework's Biotechnology Programme, August 1, 1997 to December 31, 1999. Berlin: Botanic Garden and Botanical Museum Berlin-Dahlem, Department of Biodiversity Informatics. Http://www.bgbm.org/biocise/Publications/Results/Default.htm/. Accessed 10 August 2006.

Berendsohn, W. G. (2003a). TDWG Subgroup on biological collection data: Software for biological collection management. Http://www.bgbm.org/TDWG/acc/software .htm/. Accessed 9 January 2004.

Berendsohn, W. G. (2003b). ENHSIN in the context of the evolving global biological collections information system. In *ENHSIN: The European Natural History Specimen Information Network,* ed. M. J. Scoble, 21–32. London, Natural History Museum. Http://www.nhm.ac.uk/research-curation/projects/ENHSIN/documents/assets/ ch2.pdf/. Accessed 15 December 2006.

Berg, M. (1997). *Rationalizing Medical Work: Decision-Support Techniques and Medical Practices.* Cambridge, Mass.: MIT Press.

Berners-Lee, T. (2000). *Weaving the Web: The Past, Present, and Future of the World Wide Web.* London: Texere.

Bierly, E. W. (1988). The World Climate Program: Collaboration and communication on a global scale. *Annals of the American Academy of Political and Social Science* 495 (Telescience: Scientific Communication in the Information Age): 106–116.

BioCASE secretariat (2002). BioCASE: A biological collection access service for Europe. Http://www.biocase.org/default.shtml/. Accessed 23 December 2003.

BIOSIS (n.d.). Zoological record. Http://www.biosis.org.uk/products_services/ zoorecord.html/. Accessed 31 December 2003.

Bisby, F. A. (1986). Plans for an International Legume Database (ILD). In *Lathyrus and Lathyrism: Proceedings of the International Symposium Sponsored by the Institut de Biocenotique Experimentale des Agrosystemes (IBEAS), Universite de Pau at de Pays de l'Adour, 9–13 September, 1985,* ed. A. K. Kaul and D. Combes, 22–24. New York: Third World Medical Research Foundation.

Bisby, F. A. (1993). Species-diversity knowledge systems: The Ildis prototype for Legumes. *Biotechnology R&D Trends* 700: 159–164.

Bisby, F. A., and D. L. Hawksworth (1991). What must be done to save systematics? In *Improving the Stability of Names: Needs and Options,* ed. D. L. Hawksworth, 323–326. Regnum Vegetabile: 23. Königstein: Koeltz Scientific Books.

Bisby, F. A., G. T. Prance, and P. Henderson (1998). Application for a BBSRC research grant: Investigating a federated architecture for creating a catalogue of life. Http://www.systematics.rdg.ac.uk/spice/SPICEformReading.pdf/. Accessed 29 December 2004.

Bisby, F. A., J. Shimura, M. Ruggiero, J. Edwards, and C. Haeuser (2002). Taxonomy, at the click of a mouse. *Nature* 418(6896): 367–367.

Blackmore, S. (1996). Knowing the earth's biodiversity: Challenges for the infrastructure of systematic biology. *Science* 274(5284): 63.

Bloomfield, B. P., and T. Vurdubakis (1994). Re-presenting technology—IT consultancy reports as textual reality constructions. *Sociology—The Journal of the British Sociological Association* 28(2): 455–477.

Blume, S. (1992). Whatever happened to the string and sealing wax? In *Invisible Connections: Instruments, Institutions and Science,* ed. R. Bud and S. E. Cozzens, 87–101. Bellingham, Wash.: SPIE Optical Engineering Press.

Bohlin, I. (2004). Communication regimes in competition. *Social Studies of Science* 34(3): 365–391.

Borgman, C. L. (2000). *From Gutenberg to the Global Information Infrastructure: Access to Information in the Networked World.* Cambridge, Mass.: MIT Press.

Bossert, W. (1969). Computer techniques in systematics. In *Systematic Biology* (proceedings of an international conference held at the University of Michigan, Ann Arbor, Michigan, June 14–16, 1967, sponsored by the National Research Council), 595–605. Washington, DC: National Academy of Sciences.

Bowker, G. (2000). Biodiversity datadiversity. *Social Studies of Science* 30(5): 643–684.

Bowker, G. C., and S. L. Star (1999). *Sorting Things Out: Classification and Its Consequences.* Cambridge, Mass.: MIT Press.

Bridson, D., and L. Forman (eds.) (1998). *The Herbarium Handbook.* Kew: Royal Botanic Gardens.

Brown, C. M. (2003). The changing face of scientific discourse: Analysis of genomic and proteomic database usage and acceptance. *Journal of the American Society for Information Science and Technology* 54(10): 926–938.

Brown, L. R. (1992). The Emerson Museum. *Representations* 40 (Special Issue: Seeing Science): 57–80.

Brown, N., and M. Michael (2003). A sociology of expectations: Retrospecting prospects and prospecting retrospects. *Technology Analysis and Strategic Management* 15(1): 3–18.

Bud, R., and S. E. Cozzens (1992). Introduction. In *Invisible Connections: Instruments, Institutions, and Science,* ed. R. Bud and S. E. Cozzens, xi–xiv. Bellingham, Wash.: SPIE Optical Engineering Press.

Burawoy, M. (ed.) (2000). *Global Ethnography: Forces, Connections, and Imaginations in a Postmodern World.* Berkeley: University of California Press.

Caldas, A. (2006). On web structure and digital knowledge bases: Online and offline connections in science. In *New Infrastructures for Knowledge Production: Understanding E-Science,* ed. C. Hine, 206–232. Hershey, Penn: Information Science Publishing.

Callon, M. (1987). Society in the making: The study of technology as a tool for sociological analysis. In *The Social Construction of Technological Systems: New Directions in the Sociology and History of Technology,* ed. W. E. Bijker, T. P. Hughes, and T. Pinch, 83–103. Cambridge, Mass.: MIT Press.

Cambrosio, A., P. Keating, and A. Mogoutov (2004). Mapping collaborative work and innovation in biomedicine: A computer-assisted analysis of antibody reagent workshops. *Social Studies of Science* 34(3): 325–364.

Casey, K. (2003). Issues of electronic data access in biodiversity. In *Promise and Practice in Data Sharing,* ed. P. Wouters and P. Schröder, 41–64. Amsterdam: NIWI-KNAW. Http://www.virtualknowledgestudio.nl/staff/anne-beaulieu/documents/the-public-domain-of.pdf/. Accessed 15 December 2006.

Castells, M. (2000). *The Rise of the Network Society.* Oxford: Blackwell.

Ceruzzi, P. E. (2003). *A History of Modern Computing.* Cambridge, Mass.: MIT Press.

Chelsea Physic Garden (2003). Chelsea Physic Garden. Http://www.chelseaphysic garden.co.uk/. Accessed 20 November 2003.

Chowdhury, G. G., and S. Chowdhury (1999). Digital library research: Major issues and trends. *Journal of Documentation* 55(4): 409–448.

Clapham, A. R., T. G. Tutin, and E. F. Warburg (1962). *Flora of the British Isles.* Cambridge: Cambridge University Press.

Clarke, A. E., and J. H. Fujimura (1992). What tools? Which jobs? Why right? In *The Right Tools for the Job: at Work in Twentieth-Century Life Sciences,* ed. A. E. Clarke and J. H. Fujimura, 3–44. Princeton: Princeton University Press.

Coffey, A. (1999). *The Ethnographic Self: Fieldwork and the Representation of Identity.* London: Sage.

Collins, H. M. (1975). The seven sexes: A study in the sociology of a phenomenon, or the replication of experiments in physics. *Sociology* 9(2): 205–224.

Collins, H. M. (1985). *Changing Order: Replication and Induction in Scientific Practice.* London: Sage.

Commission of the European Communities (2000). Towards a European Research Area. Http://europa.eu.int/comm/research/era/pdf/com2000-6-en.pdf/. Accessed 10 August 2006.

Conference on Open Access to Knowledge in the Sciences and Humanities (2003). Berlin Declaration on Open Access to Knowledge in the Sciences and Humanities. Http://www.zim.mpg.de/openaccess-berlin/berlin_declaration.pdf/. Accessed 10 August 2005.

Consortium for the Barcode of Life (2005). Library and laboratory: The marriage of research, data, and taxonomic literature. Draft press release. Http://barcoding.si.edu/LibraryAndLaboratory/pressrelease_draft.pdf. Accessed 10 August 2005.

Cooper, G., and S. Woolgar (1996). The research process: Context, autonomy, and audience. In *Methodological Imaginations,* ed. J. Busfield and S. Lyon, 147–163. London: Macmillan.

Coopersmith, J. (2001). Texas politics and the fax revolution. In *Information Technology and Organizational Transformation: History, Rhetoric, and Practice,* ed. J. Yates and J. van Maanen, 59–85. Thousand Oaks, Calif.: Sage.

Council of Heads of Australian Herbaria (2003). Australia's virtual herbarium. Http://www.chah.gov.au/avh/. Accessed 17 February 2004.

Crawford, E., T. Shinn, and S. Sörlin (1992). The nationalization and denationalization of the sciences: An introductory essay. In *Denationalizing Science: the Contexts of International Scientific Practice,* ed. E. Crawford, T. Shinn, and S. Sörlin, 1–42. Sociology of the Sciences Yearbook: 16. Dordrecht: Kluwer Academic.

Cronquist, A. (1968). The relevance of the national herbaria to modern taxonomic research in the United States of America. In *Modern Methods in Plant Taxonomy,* ed. V. H. Heywood, 15–22. Botanical Society of the British Isles Conference Report No. 10. London: Academic Press.

Daston, L. (2004). Type specimens and scientific memory. *Critical Inquiry* 31(1): 153–182.

Daston, L., and P. Galison (1992). The image of objectivity. *Representations* 40 (Special Issue: Seeing Science): 81–128.

David, M., and D. Zeitlyn (1996). What are they doing? Dilemmas in analyzing bibliographic searching: Cultural and technical networks in academic life. *Sociological*

Research Online 1(4). Http://www.socresonline.org.uk/1/4/2.html/. Accessed 15 December 2006.

Davidson, T., R. Sooryamoorthy, and W. Shrum (2002). Kerala connections: Will the Internet affect science in developing areas? In *The Internet in Everyday Life,* ed. B. Wellman and C. Haythornthwaite, 496–519. Malden, Mass.: Blackwell.

de Laet, M., and A. Mol (2000). The Zimbabwe bush pump: Mechanics of a fluid technology. *Social Studies of Science* 30(2): 225–264.

Derrida, J. (1998). *Archive Fever: A Freudian Impression.* Chicago: University of Chicago Press.

Dodge, M., and R. Kitchin (2000). *Mapping Cyberspace.* London: Routledge.

Dowling, D. (1999). Experimenting on theories. *Science in Context* 12(2): 261–273.

Dyck, N. (2000). Home field advantage. In *Constructing the Field: Ethnographic Fieldwork in the Contemporary World,* ed. V. Amit, 32–53. London: Routledge.

Edwards, J. L., M. A. Lane, and E. S. Nielsen (2000). Interoperability of biodiversity databases: Biodiversity information on every desktop. *Science* 289(5488): 2312.

Elvebakk, B. (2006). Networks of objects: Practical preconditions for electronic communication. In *New Infrastructures for Knowledge Production: Understanding E-Science,* ed. C. Hine, 120–142. Hershey, Penn.: Information Science Publishing.

Embury, S. M., S. M. Brandt, J. S. Robinson, I. Sutherland, F. A. Bisby, W. A. Gray, A. C. Jones, and R. J. White (2001). Adapting integrity enforcement techniques for data reconciliation. *Information Systems* 26(8): 657–689.

Entomology Libraries and Information Network (2000). ELIN Conference speakers' abstracts. Http://www.nhm.ac.uk/hosted_sites/elin/Elin_7.htm/. Accessed 10 August 2005.

Erwin, T. L. (1997). Biodiversity at its utmost: Tropical forest beetles. In *Biodiversity II,* ed. M. L. Reaka-Kudla, D. E. Wilson, and E. O. Wilson, 27–40. Washington, DC: Joseph Henry Press.

ETAN Expert Working Group (1999). Transforming European science through information and communication technologies: Challenges and opportunities of the digital age. European Commission. Ftp://ftp.cordis.europa.eu/pub/etan/docs/ict-report.pdf. Accessed 10 August 2006.

Expert Center for Taxonomic Information (n.d.). Expert Center for Taxonomic Information. Http://www.eti.uva.nl/. Accessed 1 December 2004.

Fairchild Tropical Garden (1999). Fairchild Tropical Garden Botanical Resource Centre and Virtual Herbarium. Http://www.virtualherbarium.org/. Accessed 22 December 2003.

Felsenstein, J. (n.d.). Phylogeny programs. Http://evolution.genetics.washington .edu/phylip/software.html/. Accessed 28 July 2005.

Finholt, T. A., and G. M. Olson (1997). From laboratories to collaboratories: A new organizational form for scientific collaboration. *Psychological Science* 8(1): 28–36.

Fog Olwig, K., and K. Hastrup (eds.) (1997). *Siting Culture: The Shifting Anthropological Object*. New York: Routledge.

Foot, K. A., S. M. Schneider, M. Dougherty, M. Xenos, and E. Larsen (2003). Analyzing linking practices: Candidate sites in the 2002 U.S. electoral web sphere. *Journal of Computer Mediated Communication* 8(4). Http://jcmc.indiana.edu/vol8/ issue4/foot.html/. Accessed 15 December 2006.

Forey, P. L., R. A. Fortey, P. Kenrick, and A. B. Smith (2004). Taxonomy and fossils: A critical appraisal. *Philosophical Transactions of the Royal Society of London Series B-Biological Sciences* 359(1444): 639–653.

Forey, P. L., C. J. Humphries, I. J. Kitching, R. W. Scotland, D. J. Siebert, and D. M. Williams (1993). *Cladistics: A Practical Course in Systematics*. Oxford: Oxford University Press.

Fry, J. (2006). Coordination and control of research practice across scientific fields: Implications for a differentiated e-science. In *New Infrastructures for Knowledge Production: Understanding E-Science,* ed. C. Hine, 167–187. Hershey, Penn.: Information Science Publishing.

Fujimura, J. H. (1987). Constructing do-able problems in cancer-research—Articulating alignment. *Social Studies of Science* 17(2): 257–293.

Fujimura, J. H. (1988). The molecular biological bandwagon in cancer-research—Where social worlds meet. *Social Problems* 35(3): 261–283.

Fujimura, J. H. (1996). *Crafting Science: A Sociohistory of the Quest for the Genetics of Cancer*. Cambridge, Mass.: Harvard University Press.

Fujimura, J. H., and M. Fortun (1996). Constructing knowledge across social worlds: The case of DNA sequence databases in molecular biology. In *Naked Science: Anthropological Inquiry into Boundaries, Power, and Knowledge,* ed. L. Nader, 160–173. New York: Routledge.

Galison, P. (1997). *Image and Logic: Material Culture of Microphysics*. Chicago: University of Chicago Press.

Gee, H. (2000). Online naming of species opens digital age for taxonomy. *Nature* 406(6810): 278.

Gewin, V. (2002). Taxonomy—All living things, online. *Nature* 418(6896): 362–363.

Gieryn, T. (1999). *Cultural Boundaries of Science: Credibility on the Line*. Chicago: University of Chicago Press.

Gieryn, T. F. (1998). Biotechnology's private parts (and some public ones). In *Making Space for Science: Territorial Themes in the Shaping of Knowledge,* ed. C. Smith and J. Agar, 281–312. Basingstoke: Macmillan.

Gillespie, A. E., and R. Richardson (2000). Teleworking and the city: Myths of workplace transcendence and travel reduction. In *Cities in the Telecommunications Age: The Fracturing of Geographies,* ed. J. O. Wheeler, Y. Aoyama, and B. Wharf, 228–245. London: Routledge.

Giulini, P. (1995). Dendrological collections. In *The Botanical Garden of Padua 1545–1995,* ed. A. Minelli, 260–262. Venice: Marsilio Editori.

Glasner, P., H. Rothman, and D. Travis (1995). Exploring organisational issues in British genomic research. *Genetic Engineer and Biotechnologist* 15(2,3): 125–134.

Glasner, P., H. Rothman, and W. C. Yee (1998). The UK Human Genome Mapping Resource Centre: A user analysis. In *Social Management of Genetic Engineering,* ed. P. Wheale, R. von Schomberg, and P. Glasner, 63–75. Aldershot: Gower.

Global Biodiversity Information Facility (2000). Memorandum of Understanding for the Global Biodiversity Information Facility. Http://www.gbif.org/GBIF_org/documents/mou_html/. Accessed 3 December 2004.

Godfray, H. C. J. (2002a). How might more systematics be funded? *Antenna* 26(1): 11–17. Http://www.ucl.ac.uk/taxome/godfray02.doc/. Accessed 15 December 2006.

Godfray, H. C. J. (2002b). Challenges for taxonomy—The discipline will have to reinvent itself if it is to survive and flourish. *Nature* 417(6884): 17–19.

Godfray, H. C. J. (2002c). Towards taxonomy's "glorious revolution." *Nature* 420(6915): 461–461.

Godfray, H. C. J., and S. Knapp (2004). Taxonomy for the twenty-first century—Introduction. *Philosophical Transactions of the Royal Society of London Series B-Biological Sciences* 359(1444): 559–569.

Gokalp, I. (1990). Turbulent reactions: Impact of new instrumentation on a borderland scientific domain. *Science, Technology, and Human Values* 15(3): 284–304.

Google (2004). Google web search features. Http://www.google.com/help/features.html/. Accessed 15 December 2006.

Groenewegen, P., and P. Wouters (2004). Genomics, ICT, and the formation of R&D networks. *New Genetics and Society* 23(2): 167–185.

Guala, G. F. (2000). Lessons from the virtual herbarium. Http://www.virtualherbarium.org/Lessons%20from%20the%20Virtual%20Herbarium_files/frame.htm/. Accessed 22 December 2003.

Guice, J. (1999). Designing the future: The culture of new trends in science and technology. *Research Policy* 28(1): 81–98.

Gunn, S. (2003). Take one plant . . . *Kew* (autumn 2003): 24–25.

Gupta, A., and J. Ferguson (eds.) (1997). *Anthropological Locations: Boundaries and Grounds of a Field Science.* Berkeley: University of California Press.

Haas, F. (2006). Taxonomists and users: Two species that rarely meet. Http://www.gti-kontaktstelle.de/TaxUsers.html/. Accessed 10 August 2006.

Hagen, J. (2001). The introduction of computers into systematic research in the United States during the 1960s. *Studies in the History and Philosophy of the Biological and Biomedical Sciences* 32(2): 291–314.

Hagen, J. (2003). The statistical frame of mind in systematic biology from quantitative zoology to biometry. *Journal of the History of Biology* 36(2): 353–384.

Hannerz, U. (1992). *Cultural Complexity: Studies in the Social Organization of Meaning.* New York: Columbia University Press.

Hansard (2002). *Sustainable Development.* House of Lords debate (2001–2002) 637. See http://www.publications.parliament.uk/pa/cm/cmhansrd.htm/.

Harries, G., D. Wilkinson, L. Price, R. Fairclough, and M. Thelwall (2004). Hyperlinks as a data source for science mapping. *Journal of Information Science* 30(5): 436–447.

Haveliwala, T. H., A. Gionis, D. Klein, and P. Indyk (2002). Evaluating strategies for similarity search on the web. In *Proceedings of the Eleventh International World Wide Web Conference, May 7–11, 2002,* ed. D. Lassner, D. De Roure, and A. Iyengar, 432–442. New York: ACM Press.

Hawksworth, D. L. (ed.) (1991). *Improving the Stability of Names: Needs and Options.* Königstein: Koeltz Scientific Books.

Haythornthwaite, C., K. J. Lunsford, G. C. Bowker, and B. C. Bruce (2006). Challenges for research and practice in distributed, interdisciplinary collaboration. In *New Infrastructures for Knowledge Production: Understanding E-Science,* ed. C. Hine, 143–166. Hershey, Penn.: Information Science Publishing.

Heath, D. (1998). Locating genetic knowledge: Picturing Marfan syndrome and its traveling constituencies. *Science, Technology, and Human Values* 23(1): 71–97.

Heath, D., E. Koch, B. Ley, and M. Montoya (1999). Nodes and queries—Linking locations in networked fields of inquiry. *American Behavioral Scientist* 43(3): 450–463.

Hedgecoe, A. M. (2003). Terminology and the construction of scientific disciplines: The case of pharmacogenomics. *Science, Technology, and Human Values* 28(4): 513–537.

Hert, P. (1997). The dynamics of on-line interaction in a scholarly debate. *Information Society* 13(4): 329–360.

Hert, P. (2003). Quasi-oralité de l'ecriture électronique et sentiment de communauté dans les débats scientifiques en ligne. *Resaux* 17(97): 211–260. Http://archivesic.ccsd .cnrs.fr/documents/archives0/00/00/05/17/. Accessed 15 December 2006.

Hess, D. J. (2001). Ethnography and the development of science and technology studies. In *Sage Handbook of Ethnography,* ed. P. Atkinson, A. Coffey, S. Delamont, J. Lofland, and L. Lofland, 234–245. Thousand Oaks, Calif.: Sage.

Hey, T. (2006). Foreword. In *New Infrastructures for Knowledge Production: Understanding E-Science,* ed. C. Hine, vi–vii. Hershey, Penn.: Information Science Publishing.

Hey, T., and A. E. Trefethen (2002). The UK e-Science core programme and the Grid. *Future Generation Computing Systems* 18(8): 1017–1031.

Heywood, V. (2001). Floristics and monography—An uncertain future? *Taxon* 50(2): 361–380.

Heywood, V. H. (1976). *Plant Taxonomy.* Institute of Biology's Studies in Biology No. 5. London: Edward Arnold.

Hilgartner, S. (1995). Biomolecular databases: New communication regimes for biology. *Science Communication* 17(2): 240–263.

Hilgartner, S., and S. I. Brandt Rauf (1994). Data access, ownership, and control—Toward empirical studies of access practices. *Knowledge-Creation Diffusion Utilization* 15(4): 355–372.

Hine, C. (1991). Nomenclatural instability and its implications for biological data processing. DPhil thesis. Department of Biology. York, University of York.

Hine, C. (1995). Representations of information technology in disciplinary development—Disappearing plants and invisible networks. *Science, Technology, and Human Values* 20(1): 65–85.

Hine, C. (2000). *Virtual Ethnography.* London: Sage.

Hine, C. (2001). Ethnography in the laboratory. In *Inside Organizations: Anthropologists at Work,* ed. D. Gellner and E. Hirsch, 61–76. Oxford: Berg.

Hine, C. (2002). Cyberscience and social boundaries: The implications of laboratory talk on the Internet. *Sociological Research Online* 7(2). Http://www.socresonline.org .uk/7/2/hine.html/. Accessed 15 December 2006.

Hine, C. (2006a). Computerization movements and scientific disciplines: The reflexive potential of e-science. In *New Infrastructures for Knowledge Production: Understanding E-Science,* ed. C. Hine, 26–47. Hershey, Penn.: Information Science Publishing.

Hine, C. (2006b). Databases as scientific instruments and their role in the ordering of scientific work. *Social Studies of Science* 36(2): 269–298.

Hoagland, K. E. (1996). The taxonomic impediment and the Convention on Biodiversity. *Association of Systematics Collections Newsletter* 24: 62, 66–67. Http://www.sasb.org.au/TaxImp.html/. Accessed 15 December 2006.

Hooper-Greenhill, E. (1992). *Museums and the Shaping of Knowledge*. New York: Routledge.

Horton, D. (2003). Closure of the Iowa Herbarium: The issue and the impact. Taxacom discussion list (taxacom@listserv.nhm.ku.edu). Archived at http://listserv.nhm.ku.edu/archives/taxacom.html/. Message date 16 December 2003.

Hull, D. L. (1988). *Science as a Process: An Evolutionary Account of the Social and Conceptual Development of Science*. Chicago: University of Chicago Press.

Hurd, J., C. M. Brown, J. Bartlett, P. Krietz, and G. Paris (2002). The role of "unpublished" research in the scholarly communication of scientists: Digital preprints and bioinformation databases. *Proceedings of the American Society for Information Science and Technology* 39(1): 452–453.

Huxley, J. S. (ed.) (1940). *The New Systematics*. Oxford: Clarendon Press.

Iacono, S., and R. Kling (2001). Computerization movements: The rise of the Internet and distant forms of work. In *Information Technology and Organizational Transformation: History, Rhetoric and Practice,* ed. J. Yates and J. van Maanen, 93–135. Thousand Oaks, Calif.: Sage.

International Plant Names Index (2004). How can you help? Http://www.uk.ipni.org/how_to_help.html/. Accessed 28 July 2005.

International Working Group on Taxonomic Databases (2005). GBIF/TDWG Consortium funding proposal to the Gordon & Betty Moore Foundation successful. Http://www.tdwg.org/moore_announce.html/. Accessed 28 July 2005.

Jeffrey, C. (1982). *An Introduction to Plant Taxonomy*. Cambridge: Cambridge University Press.

Jeffrey, C. (1989). *Biological Nomenclature*. London: Edward Arnold.

Joerges, B., and T. Shinn (2000). Research-technology: Instrumentation between science state and industry. In *Schriftenreihe der Forschungsgruooe "Metropolenforschung" des Forschungsschwerpunkts Technik–Arbeit–Umwelt am Wissenschaftszentrum Berlin fur Sozialforschung*. FS II(00-503). Http://bibliothek.wz-berlin.de/pdf/2000/ii00-503.pdf/. Accessed 15 December 2006.

Joerges, B., and T. Shinn (eds.) (2001). *Instrumentation between Science, State, and Industry*. Sociology of the Sciences Yearbook 22. Dordrecht: Kluwer Academic Publishers.

Jordan, K., and M. Lynch (1998). The dissemination, standardization, and routinization of a molecular biological technique. *Social Studies of Science* 28(5–6): 773–800.

Keating, P., A. Cambrosio, and M. Mackenzie (1992). The tools of the discipline: Standards, models, and measures in the affinity/avidity controversy in immunology. In *The Right Tools for the Job: At Work in Twentieth-Century Life Sciences,* ed. A. E. Clarke and J. H. Fujimura, 312–354. Princeton: Princeton University Press.

Keller, E. F. (2002). *Making Sense of Life: Explaining Biological Development with Models, Metaphors, and Machines.* Cambridge, Mass.: Harvard University Press.

Kling, R., J. Fortuna, and A. King (2001). The remarkable transformation of E-Biomed into PubMed Central. Http://www.slis.indiana.edu/CSI/WP/wp01-03B.html/. Accessed 10 August 2005.

Kling, R., and S. Iacono (1988). The mobilization of support for computerization: The role of computerization movements. *Social Problems* 35(3): 226–243.

Kling, R., and S. Iacono (1996). Computerization movements and tales of technological utopianism. In *Computerization and Controversy: Value Conflicts and Social Choices,* ed. R. Kling, 85–105. San Diego: Academic Press.

Kling, R., and G. McKim (2000). Not just a matter of time: Field differences in the shaping of electronic media in supporting scientific communication. *Journal of the American Society for Information Science* 51(14): 1306–1320.

Kling, R., L. B. Spector, and J. Fortuna (2004). The real stakes of virtual publishing: The transformation of E-Biomed into PubMed central. *Journal of the American Society for Information Science and Technology* 55(2): 127–148.

Knapp, S., R. M. Bateman, N. R. Chalmers, C. J. Humphries, P. S. Rainbow, A. B. Smith, P. D. Taylor, R. I. Vane-Wright, and M. Wilkinson (2002). Taxonomy needs evolution, not revolution—Some changes are clearly necessary, but science cannot be replaced by informatics. *Nature* 419(6907): 559.

Knapp, S., G. Lamas, E. N. Lughadha, and G. Novarino (2004). Stability or stasis in the names of organisms: The evolving codes of nomenclature. *Philosophical Transactions of the Royal Society of London Series B–Biological Sciences* 359(1444): 611–622.

Knorr-Cetina, K. (1981). *The Manufacture of Knowledge: An Essay on the Constructivist and Contextual Nature of Science.* Oxford: Pergamon Press.

Knorr-Cetina, K. (1999). *Epistemic Cultures: How the Sciences Make Knowledge.* Cambridge, Mass.: Harvard University Press.

Kouzes, R. T., J. D. Myers, and W. A. Wulf (1996). Collaboratories: Doing science on the Internet. *IEEE Computer* 29(8): 40–46.

Krell, F. T. (2000). Impact factors aren't relevant to taxonomy. *Nature* 405(6786): 507–508.

Krell, F. T. (2002). Why impact factors don't work for taxonomy. *Nature* 415(6875): 957.

Lakhani, K. R., and E. von Hippel (2003). How open source software works: "Free" user-to-user assistance. *Research Policy* 32(6): 923–943.

Latour, B. (1983). Give me a laboratory and I will raise the world. In *Science Observed,* ed. K. D. Knorr-Cetina and M. Mulkay, 141–170. London: Sage.

Latour, B. (1987). *Science in Action: How to Follow Engineers and Scientists through Society*. Cambridge, Mass.: Harvard University Press.

Latour, B. (1996). *Aramis, or The Love of Technology*. Cambridge, Mass.: Harvard University Press.

Latour, B., and S. Woolgar (1986). *Laboratory Life: The Construction of Scientific Facts*. Princeton: Princeton University Press.

Law, J., and M. Lynch (1988). Lists, field guides, and the descriptive organization of seeing—Birdwatching as an exemplary observational activity. *Human Studies* 11(2,3): 271–303.

Lawler, A. (2001). Up for the count? An odd combination of high-tech gurus and senior taxonomists is planning an ambitious—some say quixotic—effort to catalog and describe all species on Earth. *Science* 294(5543): 769.

Leander, K. M., and K. K. McKim (2003). Tracing the everyday "sitings" of adolescents on the Internet: A strategic adaptation of ethnography across online and offline spaces. *Education, Communication, and Information* 3(2): 211–240.

Lee, D., A. W. Lin, T. Hutton, T. Akiyama, S. Shinji, F. P. Lin, S. Peltier, and M. H. Ellisman (2003). Global telescience featuring IPv6 at iGrid2002. *Future Generation Computer Systems* 19(6): 1031–1039.

Lenoir, T. (1998a). Inscription practices and materialities of communication. In *Inscribing Science: Scientific Texts and the Materiality of Communication,* ed. T. Lenoir, 1–19. Stanford: Stanford University Press.

Lenoir, T. (1998b). Shaping biomedicine as an information science. In *Conference on the History and Heritage of Science Information Systems,* ed. M. E. Bowden, T. B. Hahn, and R.V. Williams, 27–45. Information Today, Inc. Http://www.stanford.edu/dept/HPS/TimLenoir/shapingbiomedicine.htm. Accessed 10 August 2006.

Lenoir, T., and C. Ross (1996). The naturalized history museum. In *The Disunity of Science: Boundaries, Contexts, and Power,* ed. P. Galison and D. J. Stump, 370–397. Stanford: Stanford University Press.

Lewenstein, B. V. (1995a). Do public electronic bulletin boards help create scientific knowledge? The cold fusion case. *Science, Technology, and Human Values* 20(2): 123–149.

Lewenstein, B. V. (1995b). From fax to facts—Communication in the cold-fusion saga. *Social Studies of Science* 25(3): 403–436.

Lievrouw, L. A., and K. Carley (1990). Changing patterns of communication among scientists in an era of "telescience." *Technology in Society* 12(4): 457–477.

Lueg, C., and D. Fisher (eds.) (2003). *From Usenet to CoWebs: Interacting with Social Information Spaces*. London: Springer-Verlag.

Lughadha, E. N. (2004). Towards a working list of all known plant species. *Philosophical Transactions of the Royal Society of London Series B–Biological Sciences* 359(1444): 681–687.

Lury, C. (2004). *Brands: The Logos of the Global Economy*. London: Routledge.

Lynch, M. (1985). *Art and Artifact in Laboratory Science: A Study of Shop Work and Shop Talk in a Research Laboratory*. London: Routledge and Kegan Paul.

Lynch, M. (1991). Laboratory space and the technological complex: An investigation of topical contextures. *Science in Context* 4(1): 51–78.

Lynch, M. (1993). *Scientific Practice and Ordinary Action: Ethnomethodology and Social Studies of Science*. Cambridge: Cambridge University Press.

Lynch, M., and K. Jordan (1995). Instructed actions in, of, and as molecular-biology. *Human Studies* 18(2,3): 227–244.

Mackenzie, D. (1996). *Knowing Machines*. Cambridge, Mass.: MIT Press.

MacLeod, N., and R. T. Patterson (1998). The role and the promise of electronic publishing in paleontology. *Palaeontologia Electronica* 1(1). Http://palaeoelectronica.org/1998_1/editor/issue1.htm/. Accessed 15 December 2006.

Makagon, D. (2000). Accidents should happen: Cultural disruption through alternative media. *Journal of Communication Inquiry* 24(4): 430–447.

Marcus, G. (1995). Ethnography in/of the world system: The emergence of multi-sited ethnography. *Annual Review of Anthropology* 24: 95–117.

Marcus, G. (1998). *Ethnography through Thick and Thin*. Princeton: Princeton University Press.

Martin, E. (1998). Anthropology and the cultural study of science. *Science, Technology, and Human Values* 23(1): 24–44.

Massey, J. R. (1974a). Collection and field preparation of specimens. In *Vascular Plant Systematics,* ed. A. E. Radford, W. C. Dickison, J. R. Massey, and C. R. Bell. New York: Harper and Row. Http://www.herbarium.unc.edu/courses/chpt18.html/. Accessed 17 August 2005.

Massey, J. R. (1974b). The herbarium. In *Vascular Plant Systematics,* ed. A. E. Radford, W. C. Dickison, J. R. Massey, and C. R. Bell. New York: Harper and Row. Http://www.herbarium.unc.edu/courses/chpt31.html/. Accessed 17 August 2005.

Matzat, U. (1998). Informal academic communication and scientific usage of Internet discussion groups. In *IRISS '98 International Conference*, Bristol, UK. Http://www.sosig.ac.uk/iriss/papers/paper19.htm/. Accessed 10 August 2006.

Matzat, U. (2001). Social networks and cooperation in electronic communities: A theoretical-empirical analysis of academic communication and Internet discussion groups. Http://www.ub.rug.nl/eldoc/dis/ppsw/u.matzat/. Accessed 22 November 2004.

Matzat, U. (2004). Academic communication and Internet discussion groups: Transfer of information or creation of social contacts? *Social Networks* 26(3): 221–255.

Maxwell, R. (1990). Information technology as a way of reducing the costs and time in the dissemination of scientific and technological information. Fourth Annual Dainton Lecture. London: British Library .

McCain, K. W. (1995). Mandating sharing—Journal policies in the natural-sciences. *Science Communication* 16(4): 403–431.

McCain, K. W. (2000). Sharing digitized research-related information on the World Wide Web. *Journal of the American Society for Information Science* 51(14): 1321–1327.

McGirr, N. (2000). *Nature's Connections: An Exploration of Natural History*. London: Natural History Museum.

McNeill, J. (1968). Regional and local herbaria. In *Modern Methods in Plant Taxonomy,* ed. V. H. Heywood, 33–44. Botanical Society of the British Isles Conference Report No. 10. London: Academic Press.

McNeill, J., T. F. Stuessy, N. J. Turland, and E. Hörandl (2005). XVII International Botanical Congress: Preliminary mail vote and report of Congress action on nomenclature proposals. *Taxon* 54(4): 1057–1064.

Meadows, J., and H.-D. Böcker (eds.) (1998). *Electronic Communication and Research in Europe: A Conference Organised by the Academia Europaea, Darmstadt, 15–17 April 1998*. Brussels: European Commission.

Merz, M. (1998). "Nobody can force you when you are across the ocean"—Face to face and e-mail exchanges between theoretical physicists. In *Making Space for Science: Territorial Themes in the Shaping of Knowledge,* ed. C. Smith and J. Agar, 313–329. Basingstoke: Macmillan.

Merz, M. (2006). Embedding digital infrastructure in epistemic culture. In *New Infrastructures for Knowledge Production: Understanding E-Science,* ed. C. Hine, 99–119. Hershey, Penn.: Information Science Publishing.

Metsger, D. A., and S. C. Byers (eds.) (1999). *Managing the Modern Herbarium: An Interdisciplinary Approach*. Washington, DC: Society for the Preservation of Natural History Collections.

Miller, D., and D. Slater (2000). *The Internet: An Ethnographic Approach*. Oxford: Berg.

Musgrave, T. (1998). *The Plant Hunters: Two Hundred Years of Adventure and Discovery around the World.* London: Ward Lock.

Myers, G. (1985). Texts as knowledge claims: The social construction of two biology articles. *Social Studies of Science* 15(4): 593–630.

Myerson, G. (1998). The electronic archive. *History of the Human Sciences* 11(4): 85–101.

Natural History Museum (2003) The 2003–07 Corporate Plan. London: The Natural History Museum. Http://www.nhm.ac.uk/info/corporate/plan/cp_final2003.pdf/. Accessed 22 December 2003.

Nentwich, M. (1999). Cyberscience: The future of research in the age of information and communication technologies (1998–002). Http://www.oeaw.ac.at/ita/cyber science.htm/. Accessed 10 August 2006.

Nentwich, M. (2001a). How online communication may affect academic knowledge production—Some preliminary hypotheses. *Internet-Zeitschrift für Kulturwissenschaften* 10. Http://www.inst.at/trans/10Nr/nentwich10.htm/. Accessed 10 August 2006.

Nentwich, M. (2001b). (Re-)De-commodification in academic knowledge distribution? *Science Studies* 14(2): 21–42.

Nentwich, M. (2003). *Cyberscience: Research in the Age of the Internet.* Vienna: Austrian Academy of Sciences Press.

Nentwich, M. (2006). Cyberinfrastructure for next generation scholarly publishing. In *New Infrastructures for Knowledge Production: Understanding E-Science,* ed. C. Hine, 189–205. Hershey, Penn.: Information Science Publishing.

Nutch, F. (1996). Gadgets, gizmos, and instruments: Science for the tinkering. *Science, Technology, and Human Values* 21(2): 214–228.

OECD (2000). The Global Research Village Conference 2000: Access to publicly financed research. Http://www.oecd.org/document/37/0,2340,en_2649_34293_1880805_1_1_1_1,00.html/. Accessed 10 August 2006.

Ogden, J., G. Walt, and L. Lush (2003). The politics of "branding" in policy transfer: The case of DOTS for tuberculosis control. *Social Science and Medicine* 57(1): 179–188.

Orgad, S. S. (2005). From online to offline and back: Moving from online to offline research relationships with informants. In *Virtual Methods: Issues in Social Research on the Internet,* ed. C. Hine, 51–66. Oxford: Berg.

Orlikowski, W. (2000). Using technology and constituting structures: A practice lens for studying technology in organizations. *Organization Science* 11(4): 404–428.

Orlikowski, W. (2001). Improvising organizational transformation over time: A situated change perspective. In *Information Technology and Organizational Transformation:*

History, Rhetoric, and Practice, ed. J. Yates and J. van Maanen, 223–274. Thousand Oaks, Calif.: Sage.

Orlikowski, W., J. Yates, K. Okamura, and M. Fujimuto (1995). Shaping electronic communication: The metastructuring of technology in the context of use. *Organization Science* 6(4): 423–444.

Oudshoorn, N., and T. Pinch (2003). Introduction: How users and non-users matter. In *How Users Matter: The Co-construction of Users and Technology,* ed. N. Oudshoorn and T. Pinch, 1–25. Cambridge, Mass.: MIT Press.

Outram, D. (1996). New spaces in natural history. In *Cultures of Natural History,* ed. N. Jardine, J. A. Secord, and E. C. Spary, 249–265. Cambridge: Cambridge University Press.

Owens, S. (2003). Future perfect. *Kew* (autumn 2003): 26–27.

Owens, S. J., and A. Prior (2000). Beset with pitfalls—Specimens and databases, intellectual property, and copyright. Presentation to *Taxonomic Databases Working Group,* Senckenberg Museum, Frankfurt, 10–12 November, 2000. Http://www.tdwg.org/tdwg2000/ipr.htm/. Accessed 10 August 2006.

Palackal, A., M. Anderson, B. P. Miller, and W. Shrum (2006). Gender stratification and e-science: Can the internet circumvent patrifocality? In *New Infrastructures for Knowledge Production: Understanding E-Science,* ed. C. Hine, 246–271. Hershey, Penn.: Information Science Publishing.

Pankhurst, R. J. (ed.) (1973). *Biological Identification with Computers.* Systematics Association Special Volume. London: Academic Press.

Pankhurst, R. J., and S. M. Walters (1971). Generation of keys by computer. In *Data Processing in Biology and Geology: Proceedings of a Symposium Held at the Department of Geology, University of Cambridge, 24–26 September 1969,* ed. J. L. Cutbill, 189–203. Systematics Association Special Volume No. 3. London: Academic Press.

Park, H. W., and M. Thelwall (2003). Hyperlink analyses of the World Wide Web: A review. *Journal of Computer Mediated Communication* 8(4). Http://jcmc.indiana.edu/vol8/issue4/park.html/. Accessed 15 December 2006.

Patterson, R. T. (2000). The secret to a long life. *Palaeontologia Electronica* 3(2). Http://palaeo-electronica.org/2000_2/toc.htm/. Accessed 15 December 2006.

Pedrotti, F. (1995). The phanerogamic herbarium. In *The Botanical Garden of Padua 1545–1995,* ed. A. Minelli, 245–259. Venice: Marsilio Editori.

Pickering, A. (ed.) (1992). *Science as Practice and Culture.* Chicago: University of Chicago Press.

Pitkin, B. (2002). Virtual collections. Darwin Centre Live presentation. Http://127.0.0.1:800/Default/www.nhm.ac.uk/darwincentre/live/pres. Accessed23 December 2003.

Poster, M. (2004). Consumption and digital commodities in the everyday. *Cultural Studies* 18(2): 409–423.

Potter, J. (1996). *Representing Reality*. London: Sage.

Rachel, J., and S. Woolgar (1995). The discursive structure of the social-technical divide—The example of information-systems development. *Sociological Review* 43(2): 251–273.

Rader, L., and C. Ison (n.d.). The legacy of mercuric chloride. Http://www-museum.unl.edu/research/botany/mercury/merc1.html/. Accessed 22 December 2003.

Reid-Henry, S. (2003). Under the microscope. Fieldwork practice and Cuba's biotechnology industry: A reflexive affair? *Singapore Journal of Tropical Geography* 24(2): 184–197.

Resource: The Council for Museums Archives and Libraries (2002). Tessa Blackstone announces £5.2 million grants for 49 museums around the country. Http://www.mla.gov.uk/news/press_article.asp?articleid=396/. Accessed 23 February 2004.

Rice, A. L. (1999). *Voyages of Discovery: Three Centuries of Natural History Exploration*. London: Scriptom Editions.

Riggs, W., and E. Vonhippel (1994). Incentives to innovate and the sources of innovation—The case of scientific instruments. *Research Policy* 23(4): 459–469.

Rogers, R. (2002). Operating issue networks on the web. *Science as Culture* 11(2): 191–214.

Rogers, R., and N. Marres (2000). Landscaping climate change: A mapping technique for understanding science and technology debates on the World Wide Web. *Public Understanding of Science* 9(2): 1–23.

Roth, B. (1993). aff. bff. cff. . . . Taxacom discussion list (taxacom@harvarda.bitnet). Archived at http://listserv.nhm.ku.edu/archives/taxacom.html/. Message date 14 November 1993.

Royal Botanic Gardens Kew (2000). Kew record of taxonomic literature. Http://www.rbgkew.org.uk/bibliographies/KR/KRHomeExt.html/. Accessed 31 December 2003.

Royal Botanic Gardens Kew (2004). Corporate Plan 2004/05–2008/09. London: Royal Botanic Gardens Kew. Http://www.kew.org/aboutus/cp2003-2008/index.html/. Accessed 10 August 2006.

Royal Botanic Gardens Kew (n.d.-a). Herbarium collections: Preserved specimens. Http://www.rbgkew.org.uk/collections/herb_specimens.html/. Accessed 5 December 2003.

Royal Botanic Gardens Kew (n.d.-b). Herbarium digitisation. Http://www.rbgkew.org.uk/data/herb_digitisation.html/. Accessed 23 December 2003.

Sandvig, C. (2003). Policy, politics, and the local Internet. *Communication Review* 6(3): 178–183.

Schaffer, S. (1989). Glass works, Newton's prisms, and the uses of experiment. In *The Uses of Experiments: Studies in the Natural Sciences,* ed. D. Gooding, T. Pinch, and S. Schaffer, 67–104. Cambridge: Cambridge University Press.

Schaffer, S. (1992). Late Victorian metrology and its instrumentation: A manufactory of Ohms. In *Invisible Connections: Instruments, Institutions, and Science,* ed. R. Bud and S. E. Cozzens, 23–56. Bellingham, Wash.: SPIE Optical Engineering Press.

Schaub, M., and C. P. Dunn (2002). vPlants: A virtual herbarium of the Chicago region. *First Monday* 7(5). Http://firstmonday.org/issues/issue7_5/schaub/. Accessed 15 December 2006.

Schneider, S. M., and K. A. Foot (2004). The web as an object of study. *New Media Society* 6(1): 114–122.

Schuh, R. T. (2000). *Biological Systematics: Principles and Applications.* Ithaca: Cornell University Press.

Scoble, M. J. (ed.) (2003). *ENHSIN: The European Natural History Specimen Information Network.* London: Natural History Museum.

Scoble, M. J. (2004). Unitary or unified taxonomy? *Philosophical Transactions of the Royal Society of London Series B–Biological Sciences* 359(1444): 699–710.

Select Committee on Science and Technology (1991). Systematic Biology Research Volume 1—Report. London: HMSO.

Select Committee on Science and Technology (2002a). What on Earth? The threat to the science underpinning conservation. London, House of Lords. Session 2001-02 3rd Report HL Paper 118(i).

Select Committee on Science and Technology (2002b). What on Earth? The threat to the science underpinning conservation. Evidence. London, House of Lords. Session 2001-02 3rd Report HL Paper 118(ii).

Sellen, A. J., and R. H. R. Harper (2002). *The Myth of the Paperless Office.* Cambridge, Mass.: MIT Press.

Shetler, S. G., J. H. Beaman, M. E. Hale, L. E. Morse, J. J. Crockett, and R. A. Creighton (1971). Pilot data processing systems for floristic information. In *Data Processing in Biology and Geology: Proceedings of a Symposium Held at the Department of Geology, University of Cambridge, 24–26 September 1969,* ed. J. L. Cutbill, 275–310. Systematics Association Special Volume No. 3. London: Academic Press.

Shinn, T., and B. Joerges (2002). The transverse science and technology culture: Dynamics and roles of research-technology. *Social Science Information* 41(2): 207–251.

Silver, D. (2000). Looking backwards, looking forwards: Cyberculture studies, 1990–2000. In *Web.Studies: Rewiring Media Studies for the Digital Age,* ed. D. Gauntlett, 19–30. London: Arnold.

Sismondo, S. (2004). *An Introduction to Science and Technology Studies*. Malden, Mass.: Blackwell.

Smith, G. F., Y. Steenkamp, R. R. Klopper, S. J. Siebert, and T. H. Arnold (2003). The price of collecting life—Overcoming the challenges involved in computerizing herbarium specimens. *Nature* 422(6930): 375–376.

Smith, V. S. (2004). Lousy lists. *Systematic Biology* 53(4): 666–668.

Stace, C. A. (1980). *Plant Taxonomy and Biosystematics*. London: Edward Arnold.

Stace, C. A. (1991). *Plant Taxonomy and Biosystematics*. Cambridge: Cambridge University Press.

Star, S. L. (1992). Craft vs. commodity, mess vs. transcendence: How the right tool became the wrong one in the case of taxidermy and natural history. In *The Right Tools for the Job: At Work in Twentieth-Century Life Sciences,* ed. A. E. Clarke and J. H. Fujimura, 257–286. Princeton: Princeton University Press.

Star, S. L., and K. Ruhleder (1996). Steps toward an ecology of infrastructure: Design and access for large information spaces. *Information Systems Research* 7(1): 111–134.

Stein, K. J. (1994). The virtual herbarium. Http://home.usit.net/~info7/plants.html/. Accessed 22 December 2003.

Stevens, P. F. (2001). Angiosperm phylogeny website. Http://www.mobot.org/MOBOT/research/APweb/. Accessed 19 November 2003.

Sun, M. (1987). Botany bids for the "big science" league; a multimillion dollar proposal to catalog the flora of North America has caused some strife among botanists. *Science* 237(4818): 967.

Sunderland, P. L. (1999). Fieldwork and the phone. *Anthropological Quarterly* 72(3): 105–117.

Taxonomic Databases Working Group (1985). Minutes of the First Meeting at the Conservatoire et Jardin Botanique, Geneva. Http://www.tdwg.org/first_minutes.pdf/. Accessed 21 December 2004.

Thackray, J., and B. Press (2001). *The Natural History Museum: Nature's Treasurehouse*. London: Natural History Museum.

Thelwall, M. (2004). *Link Analysis: An Information Science Approach*. Amsterdam: Elsevier Academic Press.

Thelwall, M. (2006). Interpreting social science link analysis research: A theoretical framework. *Journal of the American Society for Information Science and Technology* 57(1): 60–68.

Thelwall, M., and D. Wilkinson (2004). Finding similar academic web sites with links, bibliometric couplings, and colinks. *Information Processing and Management* 40(3): 515–526.

Thiele, K., and D. Yeates (2002). Tension arises from duality at the heart of taxonomy—Names must both represent a volatile hypothesis and provide a key to lasting information. *Nature* 419(6905): 337.

Traweek, S. (1988). *Beamtimes and Lifetimes: The World of High Energy Physicists*. Cambridge, Mass.: Harvard University Press.

Turkle, S. (1984). *The Second Self: Computers and the Human Spirit*. Cambridge, Mass.: MIT Press.

Turkle, S. (1995). *Life on Screen: Identity in the Age of the Internet*. New York: Simon and Schuster.

Turkle, S. (ed.) (2007). *Evocative Objects: Things We Think With*. Cambridge, Mass.: MIT Press.

UK Research Councils e-Science Core Programme (n.d.). Http://www.rcuk.ac.uk/escience/. Accessed 20 October 2003.

University of North Carolina Herbarium (n.d.). Plant Information Centre. Http://www.ibiblio.org/pic/index.htm/. Accessed 22 December 2003.

Urry, J. (2000). Mobile sociology. *British Journal of Sociology* 51(1): 185–203.

Valdecasas, A. G., S. Castroviejo, and L. F. Marcus (2000). Reliance on the citation index undermines the study of biodiversity. *Nature* 403(6771): 698–698.

van der Velde, G. (2001). Taxonomists make a name for themselves. *Nature* 414 (6860): 148.

van Helden, A., and T. L. Hankins (1994). Introduction: Instruments in the history of science. *Osiris* 9(Instruments): 1–6.

van Lente, H., and A. Rip (1998). The rise of membrane technology: From rhetorics to social reality. *Social Studies of Science* 28(2): 221–254.

Vann, K., and G. Bowker (2006). Interest in production: On the configuration of technology-bearing labours for epistemic-IT. In *New Infrastructures for Knowledge Production: Understanding E-Science,* ed. C. Hine, 71–97. Hershey, Penn.: Information Science Publishing.

Vernon, K. (1993). Desperately seeking status—Evolutionary systematics and the taxonomists' search for respectability 1940–60. *British Journal for the History of Science* 26(89): 207–227.

Walsh, J. P., and T. Bayma (1996a). Computer networks and scientific work. *Social Studies of Science* 26(3): 661–703.

Walsh, J. P., and T. Bayma (1996b). The virtual college. *Information Society* 12(4): 343–363.

Walsh, J. P., S. Kucker, M. Maloney, and S. M. Gabbay (2000). Connecting minds: CMC and scientific work. *Journal of the American Society for Information Science* 51(14): 1295–1305.

Warner, D. J. (1990). What is a scientific instrument, when did it become one, and why. *British Journal for the History of Science* 23(76): 83–93.

Watson, J. (2004). Biology and IT join to create biobanks. *Computing* (24 November). Http://www.computing.co.uk/computing/analysis/2076008/biology-join-create-biobanks/. Accessed 10 August 2005.

Webster, F. (1995). *Theories of the Information Society*. London: Routledge.

Wellcome Trust (2003). An economic analysis of scientific research publishing. London: Wellcome Trust. Http://www.wellcome.ac.uk/doc_WTD003181.html/. Accessed 10 August 2006.

Wellcome Trust (2004). Costs and business models in scientific research publishing. London: Wellcome Trust. Http://www.wellcome.ac.uk/doc_WTD003185.html/. Accessed 10 August 2006.

Wheeler, Q. D. (2004). Taxonomic triage and the poverty of phylogeny. *Philosophical Transactions of the Royal Society of London Series B–Biological Sciences* 359(1444): 571–583.

Winston, J. E. (1999). *Describing Species: Practical Taxonomic Procedure for Biologists*. New York: Columbia University Press.

Woolgar, S. (2002a). Five rules of virtuality. In *Virtual Society? Technology, Cyberbole, Reality*, ed. S. Woolgar, 1–22. Oxford: Oxford University Press.

Woolgar, S. (ed.) (2002b). *Virtual Society? Technology, Cyberbole, Reality*. Oxford: Oxford University Press.

Woolgar, S., and C. Coopmans (2006). Virtual witnessing in a virtual age: A prospectus for social studies of e-science. In *New Infrastructures for Knowledge Production: Understanding E-Science*, ed. C. Hine, 1–25. Hershey, Penn.: Information Science Publishing.

Woolgar, S., C. Coopmans, D. Neyland, and E. Simakova (2005). Does STS mean business too? A provocation piece for a one-day workshop at the Saïd Business School, University of Oxford, 29 June 2005. Http://www.sbs.ox.ac.uk/NR/rdonlyres/ EB270F97-3AF8-43F7-A79C-8D6D51176827/993/stsworkshop2.pdf/. Accessed 9 August 2006.

Wouters, P., and A. Beaulieu (2006). Imagining e-science beyond computation. In *New Infrastructures for Knowledge Production: Understanding E-Science*, ed. C. Hine, 48–70. Hershey, Penn.: Information Science Publishing.

Wulf, W. (1993). The collaboratory opportunity. *Science* 261(5123): 854–855.

Wynne, B. (1996). SSK's identity parade: Signing-up, off-and-on. *Social Studies of Science* 26(2): 357–391.

Wyse Jackson, P. (2003). Botanic gardens: Their heritage and their future. *Kew* (autumn 2003): 56.

Xu, X. B., A. C. Jones, N. Pittas, W. A. Gray, N. J. Fiddian, R. J. White, J. Robinson, F. A. Bisby, and S. M. Brandt (2001). Experiences with a hybrid implementation of a globally distributed federated database system. In *Advances in Web-Age Information Management, Proceedings,* 212–222. Lecture Notes in Computer Science: 2118.

Yates, J. (1989). *Control through Communication: The Rise of System in American Management.* Baltimore and London: Johns Hopkins University Press.

Yates, J. (1994). Evolving information use in firms, 1850–1920: Ideology and information techniques and technologies. In *Information Acumen: The Understanding and Use of Information in Modern Business,* ed. L. Bud-Frierman, 26–50. London: Routledge.

Yates, J., and J. van Maanen (2001). Introduction. In *Information Technology and Organizational Transformation: History, Rhetoric, and Practice,* ed. J. Yates and J. van Maanen, xi–xvi. Thousand Oaks, Calif.: Sage.

Zander, R. H. (2003). Flora online archives. Http://www.mobot.org/plantscience/ResBot/FO/FloraOnline.htm/. Accessed 10 August 2005.

Zander, R. H. (2004). (180-181) Report of the Special Committee on Electronic Publishing with two proposals to amend the Code. *Taxon* 53(2): 592–594.

Zeitlyn, D., M. David, and J. Bex (1999). Knowledge lost in information: Patterns of use and non-use of networked bibliographic resources. British Library Research and Innovation Centre. RIC/G/313.

Zenzen, M., and S. Restivo (1982). The mysterious morphology of immiscible liquids: A study of scientific practice. *Social Science Information* 21(3): 447–473.

Ziegler, P., and K. R. Dittrich (2004). Three decades of data integration—All problems solved? In *Proceedings of the18th IFIP World Computer Congress (WCC 2004),* vol. 12: *Building the Information Society. Toulouse, France, August 22–27, 2004,* ed. R. Jacquart, 3–12. Dordrecht: Kluwer.

Inside Technology
edited by Wiebe E. Bijker, W. Bernard Carlson, and Trevor Pinch

Inventing the Internet
Janet Abbate

Calculating a Natural World: Scientists, Engineers and Computers during the Rise of US Cold War Research
Atsushi Akera

The Languages of Edison's Light
Charles Bazerman

Rationalizing Medical Work: Decision-Support Techniques and Medical Practices
Marc Berg

Of Bicycles, Bakelites, and Bulbs: Toward a Theory of Sociotechnical Change
Wiebe E. Bijker

Shaping Technology/Building Society: Studies in Sociotechnical Change
Wiebe E. Bijker and John Law, editors

Insight and Industry: On the Dynamics of Technological Change in Medicine
Stuart S. Blume

Digitizing the News: Innovation in Online Newspapers
Pablo J. Boczkowski

Memory Practices in the Sciences
Geoffrey C. Bowker

Science on the Run: Information Management and Industrial Geophysics at Schlumberger, 1920–1940
Geoffrey C. Bowker

Sorting Things Out: Classification and Its Consequences
Geoffrey C. Bowker and Susan Leigh Star

Designing Engineers
Louis L. Bucciarelli

Artifical Experts: Social Knowledge and Intelligent Machines
H. M. Collins

The Closed World: Computers and the Politics of Discourse in Cold War America
Paul N. Edwards

Governing Molecules: The Discursive Politics of Genetic Engineering in Europe and the United States
Herbert Gottweis

Ham Radio's Technical Culture
Kristen Haring

The Radiance of France: Nuclear Power and National Identity after World War II
Gabrielle Hecht

On Line and On Paper: Visual Representations, Visual Culture, and Computer Graphics in Design Engineering
Kathryn Henderson

Systematics as Cyberscience: Computers, Change, and Continuity in Science
Christine Hine

Unbuilding Cities: Obduracy in Urban Sociotechnical Change
Anique Hommels

Pedagogy and the Practice of Science: Historical and Contemporary Perspectives
David Kaiser, editor

Biomedical Platforms: Reproducing the Normal and the Pathological in Late-Twentieth-Century Medicine
Peter Keating and Alberto Cambrosio

Constructing a Bridge: An Exploration of Engineering Culture, Design, and Research in Nineteenth-Century France and America
Eda Kranakis

Making Silicon Valley: Innovation and the Growth of High Tech, 1930–1970
Christophe Lécuyer

Viewing the Earth: The Social Construction of the Landsat Satellite System
Pamela E. Mack

Inventing Accuracy: A Historical Sociology of Nuclear Missile Guidance
Donald MacKenzie

Knowing Machines: Essays on Technical Change
Donald MacKenzie

Mechanizing Proof: Computing, Risk, and Trust
Donald MacKenzie

An Engine, Not a Camera: How Financial Models Shape Markets
Donald MacKenzie

Building the Trident Network: A Study of the Enrollment of People, Knowledge, and Machines
Maggie Mort

How Users Matter: The Co-Construction of Users and Technology
Nelly Oudshoorn and Trevor Pinch, editors

Building Genetic Medicine: Breast Cancer, Technology, and the Comparative Politics of Health Care
Shobita Parthasarathy

Framing Production: Technology, Culture, and Change in the British Bicycle Industry
Paul Rosen

Coordinating Technology: Studies in the International Standardization of Telecommunications
Susanne K. Schmidt and Raymund Werle

Structures of Scientific Collaboration
Wesley Shrum, Joel Genuth, and Ivan Chompalov

Making Parents: The Ontological Choreography of Reproductive Technology
Charis Thompson

Everyday Engineering: An Ethnography of Design and Innovation
Dominique Vinck, editor

Index